Environmental Ion Exchange

Principles and Design

Anthony M. Wachinski, Ph.D., P.E.
Director of Technology Transfer
Thames Water Utilities
The Woodlands, Texas

James E. Etzel, Ph.D., P.E.
Professor Emeritus
Purdue University
Lafayette, Indiana

Acquiring Editor:	Joel Stein
Project Editor:	Les Kaplan
Marketing Manager:	Greg Daurelle
Direct Marketing Manager:	Arline Massey
Cover design:	Denise Craig
Manufacturing:	Sheri Schwartz

Library of Congress Cataloging-in-Publication Data

Wachinski, Anthony M.
Environmental ion exchange : principles and design / Anthony M. Wachinski, James E. Etzel.
p. cm.
Includes bibliographical references and index.
ISBN 0-87371-956-5
1. Water—Purification—Ion exchange process. 2. Sewage—Purification—Ion exchange process. I. Etzel, James E. II. Title.
TD468.W22 1996
628.1'66—dc20 96-31353
CIP

International Standard Book Number 0-87371-956-5
Library of Congress Card Number 96-31353
Printed in the United States of America 1 2 3 4 5 6 7 8 9 0
Printed on acid-free paper

DEDICATION

Dr. Wachinski dedicates this book to his wife, Linda, for her sarcastic impatience and for betting her piggy-bank money of $82.45 that he would never get this done.

PREFACE

The purpose of this book is to concentrate as much ion exchange information as possible into one volume — the goal is for people to refer to the text as the "Ion Exchange Bible." This book covers all ion exchange related design and application issues and includes tables, graphs, and conversion tables. We have also provided a review of the chemistry and hydraulics required for the design process. Through our work developing material for college-level ion exchange courses, we have found that a thorough knowledge of bar graphs and lime-soda softening is necessary and are including a chapter on these topics. We also provide engineering through extensive design examples, taking the reader through the entire design process. Each chapter is a stand-alone to allow the user rapid access to the topic of interest. Many chapters provide a reference section for further reading and research.

Few books currently exist that cover ion exchange in such detail. Most ion exchange information is available either in a book chapter, in the commercial literature made available by resin manufacturing companies, or through articles and theses. During Dr. Etzel's tenure as head of the environmental engineering area at Purdue University, he was noted as the leader in the practical application of ion exchange research and development. Dr. Etzel holds six patents dealing with ion exchange; Dr. Wachinski holds three.

This book is intended as a text for undergraduate and graduate courses related to ion exchange, as a reference book on the topic, and for the engineer involved in providing cost-effective, innovative solutions to water/wastewater problems where ion exchange is an applicable alternative process for water/wastewater handling.

Anthony M. Wachinski

Table of Contents

Introduction and Historical Perspective

Ion exchange is probably one of the most widely used scientific phenomena in the world today, with applications in chemistry, biochemistry, manufacturing, soil-water interactions, medicine, and industrial and municipal water treatment, to name just a few. Specialized applications such as demineralization, deionization, industrial water treatment, and chemical and water recovery are common. By following the historical development of ion exchange, one can learn about the natural and synthetic ion exchange materials used and developed through the years — many of these naturally occurring substances are now being evaluated as throw-away materials for ion exchange because of the problems associated with the handling and disposal of ion exchange regeneration wastewaters.

Records indicate that the concept of ion exchange existed at least back to Aristotle's time and possibly biblical times when Moses sweetened the waters of Marah. A recent interpretation of the miracle suggests that oxidized cellulose of the log that Moses cast into the waters had formed carboxyl groups that removed the "bitter" Epsom salt (magnesium sulfate) by an ion exchange reaction. Thus Moses succeeded in preparing drinking water, undoubtedly by an ion exchange technique he developed on an industrial scale (*Exodus* 15:23–25).

Alchemists searched for a material (chemical) that would change seawater to drinking water long before Arrhenius proposed the existence of ions in water (circa 1887). Aristotle, circa 340 B.C., used 20 earthen vessels containing a material with ion exchange properties to produce fresh water.

In 1850 the English agricultural chemist H. S. Thompson exchanged ammonium ions for calcium ions by passing a solution of ammonium sulfate through a column of soil. When he added lime to the soil, the ammonium salt was released. In the same year, T. J. Way published an extensive report entitled "On the Power of Soils to Absorb Manure." In 1854, some 33 years before Arrhenius published his concept of ionization, Way's work was presented to the Royal Agricultural Society of London. We now know that the action of the soil's double silicates, an important function in the regeneration of base exchange, makes the process practical. Subsequent investigations not only verified Way's findings but developed new materials, in particular resinous compounds, which expanded even more the application of ion exchange.

Credit for the development of the first synthetic industrial ion exchange is usually given to the German chemists, Harm and Rupler, but the use of ion exchange to condition water is credited to another German chemist, Robert Gans.

Gans produced synthesized ion exchange minerals by fusing clay, sand, and sodium carbonate (soda ash). Gans called these products "permutits," from the Latin *permuto*, which means interchange. Today, the name "zeolite" is widely used to describe ion exchange minerals that are either processed natural clays or synthesized materials.

Around 1935, simultaneous development of sulfonated coal exchangers by Leibknecht in Germany and synthetic phenol-formaldehyde materials by Adams and Holmes in England revolutionized the field of ion exchange, because unlike the aluminosilicate zeolites, these two materials were resistant to mineral acids, making hydrogen cycle exchange feasible. Both discoveries helped lead D'Alelio to the development of sulfonated, cross-linked, polystyrene resins in 1942. These resins in one form or another are presently used almost universally.

REFERENCES

Behrman, A. S., Early History of Zeolites, *J. AWWA*, 13, 223 (1925).

D'Alelio, G. F., U.S. Patent 2,366,007 (1944).

Gans, R., German Patent 174,097 and 197,111, 1906; U.S. Patents 914,405, Mar. 9, 1909; 943,535, Dec. 14, 1909; 1,131,503, Mar. 9, 1915.

Guccione, E., *Chemical Engineering*, 73, 138 (1963).

Kunin, R. and R. J. Myers, *Ion Exchange Resins*, John Wiley & Sons, New York, 1951.

Thompson, H. S., *Journal of the Royal Agricultural Society of England*, 11, 68 (1850).

Way, J. T., Influence of Time on the Absorptive Properties of Soils, *Journal of the Royal Agricultural Society of England*, 25, 313 (1850).

Way, J. T., On the Power of Soils to Absorb Manure, *Journal of the Royal Agricultural Society of England*, 11, 313 (1850).

CHAPTER 1

Basic Concepts from General Chemistry and Fundamental Hydraulics

Students and engineers often must refresh themselves on a basic topic or equation. This chapter allows one to do this without having to look in another text. In addition, use this chapter to review bar graphs and the chemistry of lime-soda softening, discussed in Chapter 2. A basic understanding of these concepts is assumed.

BASIC CHEMISTRY REVIEW

Terminology

Nucleus: (1) made up of neutrons and protons; (2) comprises 99.9% of the mass of an atom; (3) carries a positive charge

Atomic number: number of protons in the nucleus

Atomic weight: relative weight of an atom compared with C^{12} standard

Isotope: any of two or more forms of a chemical element having the same number of protons but a different atomic weight (elements are generally a mixture of isotope atomic weights not whole numbers)

Gram atomic weight (*GAW*): quantity of an element in grams corresponding to atomic weight (e.g., 16 g of O^{16} or 12 g of C^{12})

Gram molecular weight (*GMW*): molecular weight in grams of any particular compound; contains the same number of molecules (Avogadro's number) *whatever* the compound (e.g., 6.02×10^{23} O_2 oxygen atoms = 16.0 g oxygen; 6.02×10^{23} H_2O molecules = 18.0 g H_2O; 6.02×10^{23} OH^- ions = 17.0 g OH^- ions); chief significance is in preparation of molar and molal solutions; also known as a "mole"

Avogadro's number: number of atoms in 12 g pure C^{12} = 6.02×10^{23} "entities" per mole; typical units include atoms/mole, molecules/mole, and ions/mole

Equivalent weight: the weight of a compound that contains 1 g atom of available hydrogen or its chemical equivalent

$$\text{Equivalent weight} = \frac{\text{MW}}{\text{Z}}$$

where MW is the molecular weight and Z is the net positive valence. If Z is an acid, it is equal to the number of moles H^+ obtainable from one mole of the acid; if Z is base, it is equal to the number of moles H^+ with which one mole of base will react. A *net positive valence* is the valence of the positive element times its subscript

$$NaCl = 1 \times 1 = 1$$

$$K_2CO_3 = 1 \times 2 = 2$$

$$CaSO_4 = 2 \times 1 = 2$$

$$Fe_2(SO_4)_3 = 3 \times 2 = 6$$

Molar solution: 1 GMW dissolved in enough distilled water to make 1 L of solution; generally of interest in volumetric analyses

Molarity: number of moles of solute; number of liters of solution

Molal solution: 1 GMW dissolved in 1 L of distilled water (results in a volume of more than 1 L); normally used when physical measurements (e.g., freezing point, boiling point, vapor pressure) are involved

Molality: number of moles of solute; number of kilograms of solvent

Normal solution: contains one equivalent weight of a substance per liter of solution; to prepare one liter of a normal solution of an acid or base requires enough of the compound to furnish the ions to yield 1.008 g H^+ or 17.008 g OH^- and enough distilled water to make up a volume of 1 L

Normality: relation of a solution to the normal solution (e.g., 0.1 N, 0.5 N, etc.) or N = number of equivalent weights of solute or number of liters of solution

Valency/Oxidation State and Bonds

Valency theory: in very simplistic terms, a number of electrons (corresponding to the number of protons) are arranged in orderly rings around the nucleus, with the outer ring containing the valence electrons; atoms tend to gain or lose electrons so as to assume or approach complete rings (except for the inert elements, which already have complete rings)

Valency or oxidation number/state: determined by the number of electrons an atom can acquire, release, or share with other atoms

Formulas: simple for elements with fixed valences, e.g., H^+, O, F^{-1}, etc.; more complex for elements with several oxidation states, e.g., N, Cl, Mn, etc.

Bonds: ionic — formed by transfer of electrons from one atom to another; cation — an atom that is reduced (electrons are lost and atom becomes a positively charged ion); anion — an atom that is oxidized (electrons are gained and atom becomes a negatively charged ion); covalent — identical atoms share electrons equally (e.g., Cl_2, N_2, O_2, etc.); polar covalent — heteronuclear molecules with unlike atoms share electrons unequally (H_2O)

Metals and Nonmetals

Metals: those elements that easily lose electrons (oxidize) to form positively charged ions

Nonmetals: those elements that hold electrons firmly and tend to gain electrons (reduce) to form negatively charged ions

Nomenclature

There are very few hard-and-fast rules; but, in general, the following is true. Binary compounds have the ending "-ide" (e.g., HCl is hydrogen chloride; H_2S is hydrogen sulfide; SF_6, is sulfur hexafluoride). Acids containing oxygen cause the most problem; they are generally related to the oxidation state of the element characterizing the acid. The highest oxidation state is indicated by the "-ic" suffix (e.g., sulfuric, chromic, phosphoric, etc.) and gives rise to "-ate" salts (e.g., sulfate, chromate, etc.). The next to lowest oxidation state is indicated by the "-ous" suffix (e.g., sulfurous, chromous, phosphorous, etc.) and gives rise to "-ite" salts (e.g., sulfite, etc.). If a lower oxidation state exists, it is indicated by the "hypo-" prefix, that is, "hypo . . . ous" (e.g., hypochlorous, hypophosphorous, etc.) and "hypo . . . ite" (e.g., hypochlorite, etc.). If more than three acids are known, the acid with the highest oxidation state is given the prefix "per-" (e.g., perchloric) or "per . . . ate" for salts. All "per-" acids contain an element that has an oxidation state of 7.

Gas Laws

Boyle's Law: the volume of a gas varies inversely with its pressure at a constant temperature

$$P_1V_1 = P_2V_2$$

Charles' law: the volume of gas at constant pressure varies in direct proportion to absolute (°K or °R) temperature (K = degrees Kelvin; R = degrees Rankine).

$$\frac{V_1}{T_1} = \frac{V_2}{T_1}$$

Generalized gas law:

$$PV - nRT$$

where n is the number of moles of gas in a sample, R is the universal gas constant (0.082 – atm/mol) °K, T is the temperature, P is the pressure, and V is the volume

Dalton's law of partial pressures: the total pressure of a gaseous mixture is equal to the sum of the partial pressures of the components; the total pressure of a component of a gas mixture is the pressure that component would exert if it alone occupied the entire volume.

Henry's law (the most important of all gas laws in problems involving liquids):

> The mass of a slightly soluble gas that dissolves in a definite mass of a liquid at a given temperature is very nearly directly proportional to the partial pressure of that gas. This holds for gases which do not unite chemically with the solvent. (*CRC Handbook of Chemistry and Physics*, 76th ed.)

$$C_{equil} = HP_{gas}$$

where C_{equil} is the concentration of dissolved gas at equilibrium, P_{gas} is the partial pressure of gas above liquid, and H is the Henry's law constant for gas at a given temperature

Grades of Chemicals

The authors recommend *Standard Methods for the Examination of Water and Wastewater*, 17th ed., Part 102, pp. 4–5.

Dilutions

When a solution is diluted the volume is increased and the concentration is decreased, but the total amount of solute is kept constant.

$$V_1(\text{conc})_1 - V_2(\text{conc})_2$$

If we know the value of any three of these variables, we can calculate the fourth.

Lab Analysis

Sampling

Types of samples

Grab: taken more or less instantaneously; may be analyzed immediately but not necessarily; frequency based on "engineering judgment;" many environmental engineering samples are of this type

Composite: portions of the whole are collected at regular intervals and pooled to make one large sample; if sampling period is greater than 24 hr, one *must* consider detention times; amount of each individual sample must be proportional to flow at time of collection

pH

pH: a measure of the hydrogen ion concentration present in a water or wastewater; the logarithm of the reciprocal of the hydrogen ion concentration or the negative $\log_{10}$ of the hydrogen ion concentration in a water or wastewater

$$\mathrm{pH} = \log_{10} \frac{1}{\left[\mathrm{H}^+\right]}$$

$$= \log_{10}\left[\mathrm{H}^+\right]$$

Measurement of Hydrogen Ion Concentration

Pure water dissociates to yield $[H^+] = 10^{-7}$ moles/L and $[OH^-] = 10^{-7}$ moles/L (a simplification because the hydrogen atom actually exists as the hydronium ion H_3O^+). The equilibrium equation is

$$H_2O \rightarrow H^+ + OH^-$$

$$K_w = \frac{\left[H^+\right]\left[OH^-\right]}{\left[H_2O\right]}$$

Because $[H_2O]1 = [H^+]\ [OH^-] = K_w$, at 25°C,

$$10^{-7} \times 10^{-7} = 10^{-14}$$

and

$$K_w = 10^{-14}$$

which is the ionization constant of water.

Expression of pH

Because pH is a logarithmic function, *do not* calculate or report as an average value. An "average pH" has no meaning. Usually the range of pH values is given, and sometimes the median value.

Importance of pH

pH controls many chemical reactions, including coagulation, disinfection, water softening, corrosion, biochemical reactions, and ammonia removal. It also tells the design engineer what construction materials to use. The big question in industry is "where do I put the pH meter."

Acidity

Definition: base neutralizing power

Types

Phenolphthalein acidity (CO_2 acidity) pumping all of the CO_2 you can into water, will not lower the pH below 4.3 (carbonic acid)

Mineral acid: phosphoric acid in Coca-Cola aids in getting the pH down to 2.5–3.0

Methods of Measurement

Mineral acids are measured by titration to a pH of 4.3 using methyl orange as an indicator (orange on the alkaline side to salmon pink on the acid side). Mineral acidity plus acidity due to weak acids, e.g., carbonic acid, is measured by titration of the sample to the phenolphthalein end point (pH 8.3). This is called total acidity or phenolphthalein acidity. Phenolphthalein is colorless at a pH less than 8.3 and pink or red above pH 8.3. (Note: To a theoretical chemist, a pH of 7 is considered neutral. To a water chemist, who wants to know how much free or combined CO_2 is present and how much total alkalinity is present, a pH of 7 means very little. For a water chemist the dividing point between acidity and alkalinity is *not* pH 7.0 but the "M" (or total) alkalinity (acidity) end point of 4.3.

Significance of Acidity

In environmental engineering, acidity plays a part in neutralization of waste streams, corrosivity, and biological activity in public water supplies and in sewage treatment.

Alkalinity

Alkalinity: acid neutralizing power (see Table 1-1)

Table 1-1 Alkalinity Relationships

Result of titration	Hydroxide alkalinity	Carbonate alkalinity	Bicarbonate alkalinity
P[a] = 0	0	0	MO
P < 1/2 MO[b]	0	2P	MO – 2P
P = 1/2 MO	0	2P	0
P > 1/2 MO	2P – MO	2 (MO – P)	0
P – MO	MO	0	0

[a] P, phenolphthalein.
[b] MO, methyl orange.

Types

Phenolphthalein alkalinity ("P" alkalinity): a pH of 8.3 represents the end point for "phenolphthalein" or "P" alkalinity

Methyl orange or total alkalinity "MO alkalinity": a pH of 4.3 represents "total alkalinity"; it is directly related to the amount of hydroxide, carbonate, or bicarbonate alkalinity present

More specifically,

Hydroxide anions only — Samples have pH above 10. Titration with strong acid is complete at phenolphthalein end point (8.3). Hydroxide alkalinity is equal to phenolphthalein alkalinity.

Carbonate only — Samples have pH of 8.3 or higher. Titration to phenolphthalein end point is exactly one-half of total titration to pH 4.3. Carbonate alkalinity equals total alkalinity.

Hydroxide-carbonate — Samples have high pH, usually well above 10. Titration from pH 8.3 to 4.3 represents one-half of carbonate alkalinity.

Carbonate-bicarbonate — Samples have pH greater than 8.3 and less than 11. Titration to pH 8.3 represents one-half of carbonate alkalinity.

Bicarbonate only — Samples have pH 8.3 or less (usually less). Bicarbonate alkalinity equals total alkalinity.

BASIC HYDRAULICS REVIEW

Terminology

Ideal fluid: exhibits zero viscosity and shearing forces, incompressible, with uniform velocity distributions when flowing

Real fluid: exhibits finite viscosity and shearing forces, nonuniform velocity distributions when flowing, essentially noncompressible

Steady flow: mass at any cross-section is constant

Uniform flow: depth, cross-sectional area, and other elements of flow are constant from section to section

Nonuniform flow: slope of fluid surface, cross-sectional area, and velocity are changing from section to section

Pipe flow: generally that the pipe is flowing full

Open-channel flow: a free liquid surface subject to atmospheric pressure

Laminar versus Turbulent Flow

At higher velocities eddies form, either from contact of the flowing stream with solid boundaries or from contact between layers moving at different velocities. This leads to lateral mixing called turbulent flow. The point at which a laminar boundary layer first shows turbulence is described by the Reynolds number.

The ratio of inertial flow forces to viscous forces within the fluid is given by

$$N_R = \frac{VD_\rho}{\mu} = \frac{VD}{\nu}$$

where V is the velocity, D is the equivalent diameter (inside diameter for round pipe), μ is the absolute viscosity, ρ is the fluid density, and ν is the kinematic viscosity.

Hydraulic grade line: a line connecting the points to which the liquid would rise at various places along any pipe if piezometer tubes were inserted; a measure of piezometric head at these points; in an open channel, it corresponds to the water surface (see Figure 1-1).

Energy grade line: the total energy of flow in any section with reference to some datum; the sum of the elevation head z, the pressure head P/γ, and the velocity head $V^2/2g$, remains a constant (see Figure 1-2)

Turbine

A turbine removes mechanical energy from a fluid flow and exerts a negative hem. The power used by the pump is

$$R = \gamma Qhem/E_P \text{ (SI)}$$

$$P_i = \gamma Qhem/550\ E_P$$

$$bhp = \gamma Qhem/550\ E_P \text{ (Technical English units)}$$

where E_P is the pump efficiency, dimensionless, P_i is the power input (kW), (kN m/s) or specific weight of water (kN/m^3) (lb/ft^3), Q is the capacity (m^3/s) (ft^3/s), hem is the total dynamic head (m)(ft), bhp is brake horsepower, and 550 is the conversion factor for horsepower to ft-lb/s. Pump efficiencies: 60–85%.

Equations

With the energy equations for real fluids, friction must be taken into account; this is accomplished by adding terms to the right-hand side of Bernoulli's theorem.

Bernoulli's theorem is usually written:

$$\frac{P_1}{\gamma} + \frac{V_1^2}{2g} + Z_1 = \frac{P_2}{\gamma} + \frac{V_2^2}{2g} + Z_2$$

where P_1 is the pressure at section 1, V_1 is the velocity at section 1, Z_1 is the elevation at section 1, γ is the specific weight of water, g is the acceleration due to gravity, P_2 is the pressure at section 2, V_2 is the velocity at section 2, Z_2 is the elevation at section 2.

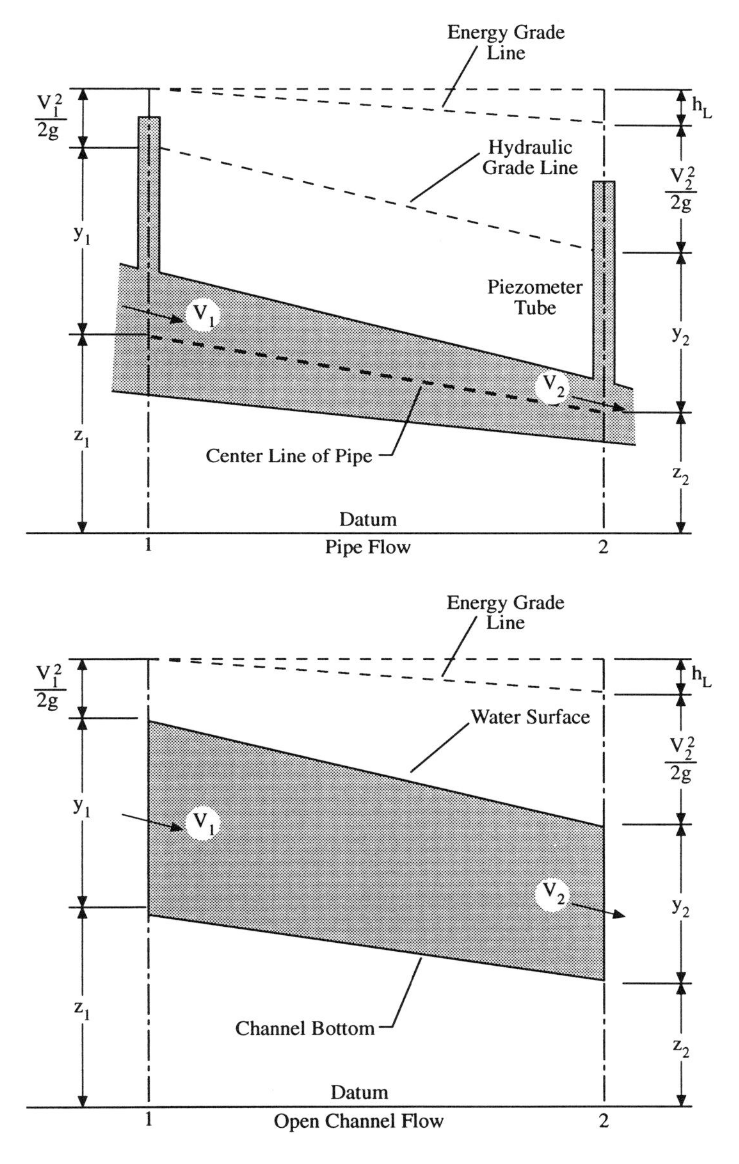

Figure 1-1 Comparison of pipe flow and open-channel flow, showing hydraulic grade line and energy grade line.

The energy equation is written

$$\frac{P_1}{\gamma} + \frac{V_1^2}{2g} + Z_1 = \frac{P_2}{\gamma} + \frac{V_2^2}{2g} + Z_2 + h_{Lf} + h_{Lm}$$

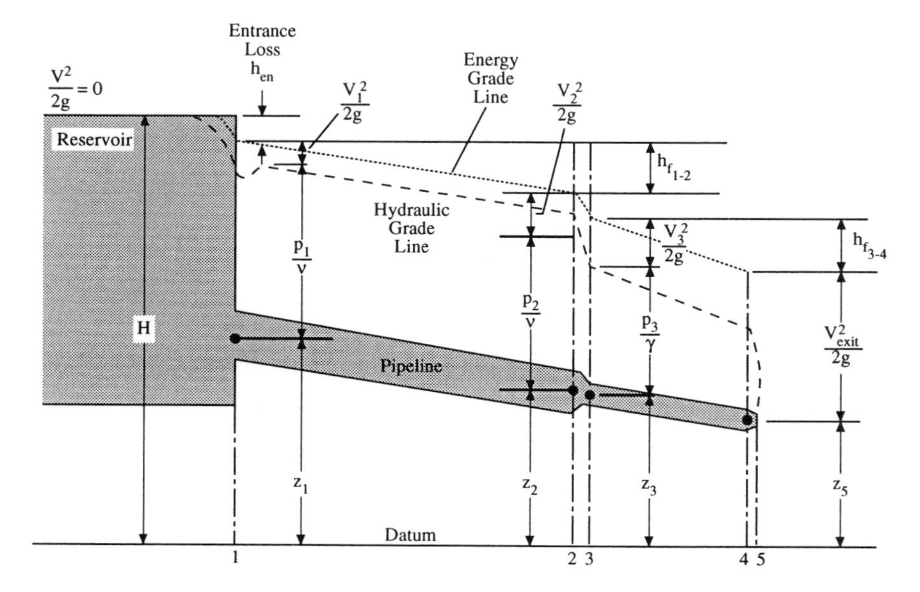

Figure 1-2 Application of energy equation to pipeline.

where h_{Lf} is the head loss due to friction and h_{Lm} is the head loss due to minor losses.

Darcy-Weisbach equation: head loss due to friction in circular pipes flowing full

$$hf = \frac{f\,LV^2}{2g}$$

or in terms of volumetric flow rate (Q)

$$h_{lf} = 8f\ LQ^2/\pi^2 gD^5$$

The hydraulic radius (R_H) is the cross-sectional flow area (A) divided by the wetted perimeter (P)

$$R_H = \frac{A}{P}$$

Hazen-Williams equation: for pipe flow involving water and wastewater force mains, pipe diameters >50 mm (2")

$$V = 0849\ CR_H^{0.63}\ S^{0.54}\ \text{(SI units)}$$

$$V = 1.318\ CR_H^{0.63}\ S^{0.54}\ \text{(TE units)}$$

For Q, when $R_H = D/4$,

$$Q = 0.278\ CD^{2.63}\ S^{0.54}\ \text{(SI units)}$$

$$Q = 0.432\ CD^{2.63}\ S^{0.54}\ \text{(TE units)}$$

Typically used Hazen-Williams coefficients include

Type of pipe	C
Pipes extremely straight and smooth	140
Pipes very smooth	130
Smooth wood, smooth masonry	120
New riveted steel, vitrified clay	110
Old cast iron, ordinary brick	100
Old riveted steel	95
Old iron in bad condition	60–80

REFERENCES

ALPHA, AWWA, WEF, Standard Methods for the Examination of Water and Wastewater, 17th ed., 1989.

Giles, R. V., Schaum's Outline Series, *Theory Fluid Mechanics and Hydraulics*, 2nd ed., McGraw-Hill, New York, 1962.

Lide, D. R., Ed., *Handbook of Chemistry and Physics*, 76th ed., CRC Press, Boca Raton, FL, 1996.

Rosenberg, J. L., *Schaum's Outline of Theory and Problems of College Chemistry*, 6th ed., *Schaum's Outline Series in Science*, McGraw-Hill, New York, 1980.

Sawyer and McCarty, *Chemistry for Environmental Engineering*, McGraw-Hill, New York, 1978, 4th ed.

SUGGESTED READING

Daugherty, R.L., et al., *Fluid Mechanics With Engineering Applications*, Eighth Edition, McGraw-Hill Book Company, New York, 1985.

CHAPTER 2

Cold Process Lime-Soda Softening

A fundamental knowledge of the chemical softening process is paramount in understanding the basic principles of ion exchange.

HARDNESS

Hardness (any material that forms a precipitate with the sodium salt of fatty acids) is classified as either permanent or temporary. The only chemical compound that forms temporary hardness is calcium bicarbonate [$Ca(HCO_3)_2$], which depends on the presence of dissolved carbon dioxide in the water to even exist. In the absence of carbon dioxide, calcium bicarbonate reverts to calcium carbonate. (*Note*: Calcium bicarbonate does not exist in the dry form.) All other hardness is permanent.

Hardness is also classified as carbonate or noncarbonate. Carbonate hardness is bicarbonate, carbonate, and hydroxide. All other hardness is noncarbonate, e.g., sulfate, nitrate, and chloride.

Powell (1954) defines hardness as "a phenomenon caused by the presence of multivalent cations (calcium and magnesium), in combination with bicarbonates, carbonates, sulfates, chlorides and nitrates." In general, iron, aluminum, and manganese can also cause water to be hard, but these substances are not ordinarily present in appreciable quantities. Carbonates are found only occasionally, in highly alkaline waters. Nitrates are usually present in small amounts and, on the average, sulfates are likely to exceed chlorides. There are, of course, many water supplies to which Shepherd's generalizations do not apply.

SOFTENING

What problems are associated with the multivalent ions, calcium and magnesium, in water warrant removal or substitution of these ions, i.e., softening?

1. The loss of CO_2 by $Ca(HCO_3)_2$ forms $CaCO_3$, which can scale pipes.
2. The reaction with soaps causes them to be less effective. It is cheaper to soften the water with lime and soda ash than to let the soap do it.
3. The formation of precipitated solids in a boiler causes a reduction in the heating efficiency of the boiler.

The removal of hardness was originally called zeolite softening, later softening or water softening was used. Common practice in the United States is to identify water softening procedures by the chemicals used in the process; thus the term lime-soda softening. The processes are further classified as either hot or cold. This chapter addresses cold lime-soda softening.

BASIC PRINCIPLES

Water softening by chemical addition uses classical precipitation methods, i.e., rapid mixing, flocculation, sedimentation, recarbonation, and filtration (if required) to remove hardness. The process converts soluble multivalent cations responsible for hardness into insoluble compounds that can be reduced to about 35 mg/L by traditional solids separation techniques without excessive use of chemicals.

Lime and soda ash can be added to obtain maximum precipitation of Ca^{++} and Mg^{++} in a sample. This has been the preferred method because it is the cheapest. Whether it will remain a viable process is yet to be decided. Unfortunately, the sludge produced by this process has little value and is difficult to dispose of. As a result the process may lose favor in the future.

The first reaction is an unwanted reaction with dissolved CO_2.

$$CO_2 + Ca(OH)_2 \rightarrow CaCO_3 + H_2O \tag{1}$$

Although a totally useless reaction, it cannot be prevented. Because of this reaction, CO_2 must be added to the water after softening. The next reactions are

$$Ca(HCO_3)_2 + Ca(OH)_2 = 2CaCO_3 + 2H_2O \tag{2}$$

$$Mg(HCO_3)_2 + 2Ca(OH)_2 = Mg(OH)_2 + 2CaCO_3 + 2H_2O \tag{3}$$

We need a method to get the lime-soda reactions to move to completion. Both Ca^{++} and Mg^{++} carbonates are soluble to some degree and thus a driving force is needed to force completion of the reaction. The usual method is to add excess chemical thus shifting the equilibrium toward the carbonate precipitates. A pH value of 10.3 is the minimum solubility point for $Mg(OH)_2$, and pH 9.3 is the minimum solubility for calcium (this minimum is 8 to 10 mg/L). Thus, to get both to a minimum some chemical trick must be used.

By adding excess lime and excess soda ash to the reaction we can lower the levels of Ca^{++} and Mg^{++} to about 8 mg/L or a total hardness of 16 mg/L as a minimum. Note that the hot process will allow you to get down to lower levels of hardness because at higher temperatures CO_2 is less soluble and thus does not load the reactions in favor of the soluble bicarbonates.

After lime-soda softening the water is supersaturated with $CaCO_3$, which will precipitate or coat piping after leaving the process. Thus we must stabilize the water before putting it out through the distribution system. The usual method for stabilization is recarbonation. In recarbonation, we add CO_2 to the water to lower the pH, thus forming bicarbonates from the carbonates and ensuring that there is no way that the materials will precipitate. All of the hydroxide and carbonate alkalinity that could combine with calcium or magnesium must be converted to bicarbonate to be sure there will not be precipitation later in the distribution.

Care must be taken in recarbonation to ensure that too much CO_2 is not introduced into the water. The required amount of CO_2 can be calculated easily from a bar diagram by calculating the amount required to destroy only the hydroxide alkalinity. Conversion of Na_2CO_3 to $NaHCO_3$ is possible but unnecessary due to the high solubility of Na_2CO_3, so it is just a waste of chemical. If all of the carbonate is converted to bicarbonate there is an excellent chance of overcarbonation with a resultant drop in pH to below 8.4. With the pH below 8.4 and excess CO_2 in the water, the water will become aggressive and cause a corrosion problem in the distribution system.

The usual method of recarbonation in all but the smallest plants is to burn methane to produce CO_2 and waste the heat. Small plants use a controlled feed of sulfuric acid to convert some of the $CaCO_3$ to $CaSO_4$, which has a higher solubility, and form H_2CO_3. [*Note*: The same reaction applies to $Mg(CO)_3$.]

The only way to reduce the total dissolved solids (TDS) of a water by precipitation is if the TDS, or part of it, is in the form of bicarbonates. The bicarbonates can be precipitated as insoluble carbonates, whereas sulfates and chlorides cannot be precipitated at any reasonable concentration.

BAR CHARTS

Bar charts are used in the field of ion exchange to make regenerant calculations and to “check” laboratory analyses. An understanding of bar charts makes the more difficult concepts associated with ion exchange easier to grasp.

An order of preference exists in nature for the combining of anions and cations. For cations the order is

Iron
Aluminum
Calcium
Magnesium
Sodium
Potassium
Hydrogen

For anions the order is

Hydroxide
Bicarbonate
Carbonate
Sulfate
Chloride
Fluoride
Nitrate

To plot a bar chart, first express all of the constituents in similar units (e.g., mg/L as $CaCO_3$ or Eq/L). Two bars are constructed equal to the scale length of total anions or cations. Plot cations on the top bar and anions on the bottom in the order shown above.

Note: In most waters, some carbon dioxide (CO_2) is dissolved in the sample but is usually not reported in the analysis. The quickest and most accurate way to obtain the CO_2 concentration is by using the TDS and pH with the chart in any recent edition of *Standard Methods for the Examination of Water and Wastewater.*

The following example illustrates how to construct a bar chart. Because both phenolphthalein (P) and methyl orange (MO) alkalinity are given, we know that the pH is greater than 8.3. (See Chapter 1 for a review of alkalinity relationships.) The carbonate ($CO_3^=$) alkalinity is 140 mg/L expressed as $CaCO_3$ and the bicarbonate alkalinity is 235 mg/L expressed as $CaCO_3$.

EXAMPLE PROBLEM: BAR CHART

Given:

			Atomic weight	Equivalent atomic weight
P alk=	70 mg/L as $CaCO_3$	Ca	40	20
MO alk =	375 mg/L as $CaCO_3$	Mg	24	12
Ca^{++} =	80 mg/L as Ca^{++}	N_d	23	23
Mg^{++} =	24 mg/L as Mg^{++}	K	39	39
Na^+ =	30 mg/L as Na^+	SO_4	96	48
K^+ =	27 mg/L as K^+	Cl	35.5	35.5
SO_4 =	14.4 mg/L as $SO_4^=$	CO_3	60	30
Cl^- =	7.1 mg/L as Cl^-			

Find: The hypothetical combinations of compounds and prepare a bar chart.

SOLUTION

$$\text{Because} \quad P < 1/2\,MO,\ CO_3^{=} = 2P = (2)(70) = 140 \text{ mg/L as } CaCO_3$$

$$HCO_3^- = T - 2P = 375 - (2)(70) = 375 - 140 = 235$$

$$= 235 \text{ mg/L as } CaCO_3$$

where T is the total alkalinity or methyl orange alkalinity.
Express constituents as calcium carbonate ($CaCO_3$)

$$Ca = 80 \times \frac{50}{20} = 200 \text{ mg/L as } CaCO_3$$

$$Mg = 24 \times \frac{50}{12} = 100 \text{ mg/L as } CaCO_3$$

$$Na = 30 \times \frac{50}{23} = 65 \text{ mg/L as } CaCO_3$$

$$K = 27 \times \frac{50}{39} = 35 \text{ mg/L as } CaCO_3$$

$$SO_4^= = 14.4 \times \frac{50}{48} = 15 \text{ mg/L as } CaCO_3$$

$$Cl^- = 7.1 \times \frac{50}{35.5} = 10 \text{ mg/L as } CaCO_3$$

Figure 2-1 shows the final bar chart.

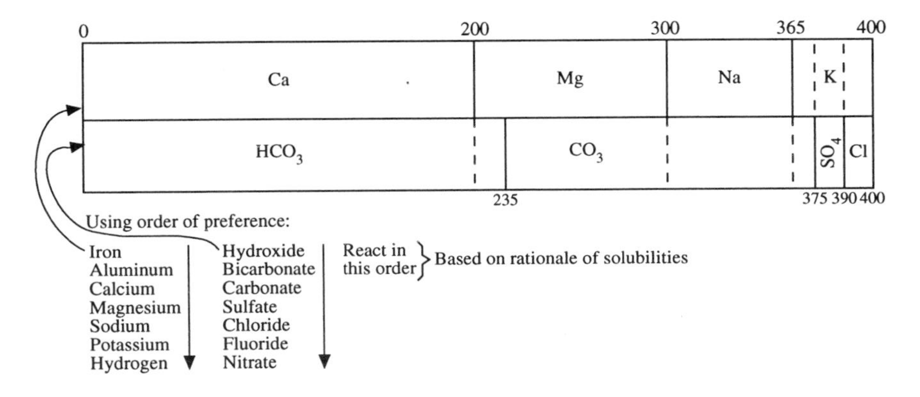

Figure 2-1 Bar chart example.

PROCESS EQUATIONS

The following equations summarize the lime-soda softening computations. Before softening can occur, lime will be consumed by any free carbon dioxide present in the water:

$$2CO_2 + Ca(OH)_2 \rightarrow Ca(HCO_3)_2 \tag{4}$$

Carbon dioxide + Calcium hydroxide (hydrated lime) → Calcium bicarbonate

This first reaction produces calcium bicarbonate, which adds hardness.

The most important reaction in water softening is the conversion of soluble calcium bicarbonate to the relatively insoluble carbonate, which can be removed by settling and filtration. Sufficient lime is added to react with the $Ca(HCO_3)_2$, which precipitates the carbonate according to the following reaction:

$$Ca(HCO_3)_2 + Ca(OH)_2 \rightarrow 2CaCO_3\downarrow + 2H_2O \tag{5}$$

Calcium bicarbonate + Calcium hydroxide (hydrated lime) → Calcium carbonate + (Water)

Partial softening may be accomplished where complete softening is not justified by adding enough lime to convert only part of the calcium bicarbonate to carbonate. Of the calcium carbonate thus formed, some will remain in solution (according to its solubility), and the remainder will precipitate.

If magnesium or sodium bicarbonate is present, lime consumption for complete treatment is shown by the following family of reactions:

$$Mg(HCO_3)_2 + Ca(OH)_2 \rightarrow MgCO_3\downarrow + CaCO_3\downarrow + 2H_2O \tag{6}$$

Magnesium bicarbonate + Calcium hydroxide (hydrated lime) → Magnesium carbonate + Calcium carbonate + (Water)

$$2NaHCO_3 + Ca(OH)_2 \rightarrow Na_2CO_3 + CaCO_3\downarrow + 2H_2O \tag{7}$$

Sodium bicarbonate + Calcium hydroxide (hydrated lime) → Sodium carbonate + Calcium carbonate + (Water)

Sodium and magnesium carbonates formed by the foregoing reaction are soluble and are not precipitated. Therefore, the additional lime used does not directly remove an equivalent amount of hardness. However, the soluble carbonates that are formed improve the performance of the process because of the excess of soluble carbonates produced.

Lime precipitates magnesium hydroxide:

$$MgCO_3 + Ca(OH)_2 \rightarrow Mg(OH)_2\downarrow + CaCO_3\downarrow$$

Magnesium carbonate + Lime → Magnesium hydroxide + Calcium carbonate (8)

$$MgSO_4 + Ca(OH)_2 \rightarrow Mg(OH)_2\downarrow + CaSO_4$$

Magnesium sulfate + Lime → Magnesium hydroxide + Calcium sulfate (9)

$$MgCl_2 + Ca(OH)_2 \rightarrow Mg(OH)_2\downarrow + CaCl_2$$

Magnesium chloride + Lime → Magnesium hydroxide + Calcium chloride (10)

Soluble magnesium carbonate uses one equivalent of lime in its formation from magnesium bicarbonate (see Equation 5) and requires a second equivalent for its conversion to the hydroxide. Magnesium sulfate and magnesium chloride, when precipitated with lime, produce a corresponding calcium salt, which adds to the calcium noncarbonate hardness of the water (Equations 8 and 9).

In general, magnesium hardness can be left in solution or removed by adding more lime, according to the requirements for the finished water. However, when lime is fed for the purpose of removing calcium bicarbonate hardness, some magnesium precipitates, and, as it consumes equivalent lime, the dosage must be increased to meet this demand. Such "incidental" magnesium removal may vary from 10% of the initial magnesium to much higher amounts where the total lime dosage is heavy.

The fourth fundamental family of reactions in lime-soda softening is the conversion by soda ash of soluble calcium sulfate and calcium chloride to insoluble calcium carbonate.

$$CaSO_4 + Na_2CO_3 \rightarrow CaCO_3\downarrow + Na_2SO_4 \quad (11)$$

Calcium sulfate + Soda ash → Calcium carbonate + Sodium sulfate

$$CaCl_2 + Na_2CO_3 \rightarrow CaCO_3\downarrow + 2NaCl \quad (12)$$

Calcium chloride + Soda ash → Calcium carbonate + Sodium chloride

The calcium carbonate thus formed precipitates along with that produced from calcium bicarbonate.

Soda ash is added not only to reduce the calcium noncarbonate hardness but also to remove magnesium noncarbonate hardness. Magnesium sulfate and chloride, upon addition of lime, form an equivalent amount of calcium noncarbonate hardness, which must then be precipitated with soda ash. In other words, the soda ash dosage is based on the total amount of noncarbonate hardness, and no distinction need be made between calcium and magnesium salts in estimating treatment requirements.

Four fundamental processes are thus involved in softening with lime and soda ash, resulting in the formation of calcium carbonate and magnesium hydroxide. The final hardness of the water will be governed by the solubility of these compounds under the existing temperature conditions and other substances present in the finished or treated water.

EXAMPLE PROBLEM: LIME SODA SOFTENING

Given:

pH = 7.1	
CO_2 = 11 mg/L	Alk = 340 mg/L as $CaCO_3$
Ca^{++} = 80 mg/L as Ca	SO_4 = 106 mg/L as $SO_4^=$
Mg^{++} = 48 mg/L as Mg	Cl^- = 36 Mg/L as Cl^-

Find: Treat 4 million gallons per day (MGD) to a residual hardness of 12 mg/L of Ca as $CaCO_3$ and 12 mg/L of Mg as $CaCO_3$.

SOLUTION

1. In terms of $CaCO_3$:

Ca^{++}	= 80 × 50/20	= 200 mg/L
Mg^{++}	= 48 × 50/12	= 200 mg/L
Na^+	= 46 × 50/23	= 100 mg/L
Alk		= 340 mg/L
$SO_4^=$	= 106 × 50/48	= 110 mg/L
Cl^-	= 36 × 50/35.5	= 50 mg/L

2. Bar graph

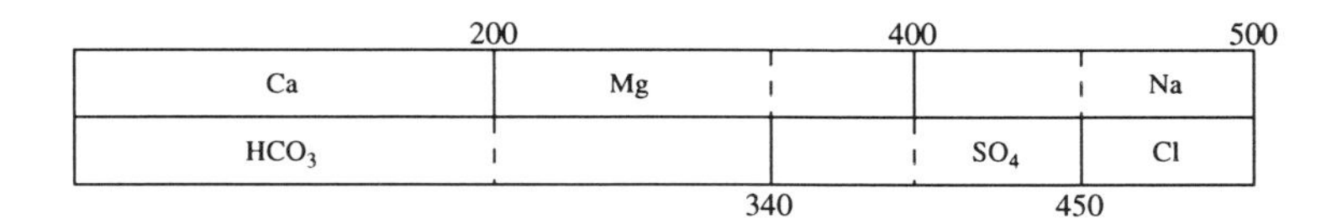

Species:

$Ca(HCO_3)_2$ = 200 mg/L as $CaCO_3$
$Mg(HCO_3)_2$ = 140 mg/L as $CaCO_3$
$MgSO_4$ = 60 mg/L as $CaCO_3$

Na_2SO_4 = 50 mg/L as $CaCO_3$
NaCl = 50 mg/L as $CaCO_3$

Chemicals:

1. $CO_2 + Ca(OH)_2 \rightarrow CaCO_3 + H_2O$
   Set up proportions:

$$\frac{44}{11} = \frac{74}{L_1} \qquad \frac{44}{11} = \frac{100}{S_1}$$

$$L_1 = \frac{(11)(74)}{44} \qquad S_1 = \frac{(100)(11)}{44}$$

$$L_1 = 18.5 \text{ mg/L} \qquad S_1 = 25 \text{ mg/L}$$

2. $Ca(HCO_3)_2 + Ca(OH)_2 \rightarrow 2CaCO_3\downarrow + 2H_2O$

$$\frac{100}{200} \qquad \frac{74}{L_2} \qquad \frac{2 \times 100}{S_2}$$

$$L_2 = 148 \text{ mg/L} \qquad S_2 = 400 \text{ mg/L}$$

In this problem, mol wt of $Ca(OH)_2$ and $CaCO_3$ are given as $Ca(OH)_2$ and $CaCO_3$. The concentration of $Ca(HCO_3)_2$ is given as $CaCO_3$. So we have to change the 200 mg/L as $CaCO_3$ to be expressed as $Ca(HCO_3)_2$, i.e., multiply by $\frac{EW}{50}$. This part of the proportion would be

$$\frac{162 \text{ (mol wt)}}{200 \times \frac{81}{50}} = \frac{74}{L_2}$$

but $162 \times \frac{50}{81} = 100 \Rightarrow$ will always be 100.

3. $Mg(HCO_3)_2 + 2Ca(OH)_2 \rightarrow 2CaCO_3 + Mg(OH)_2 + 2H_2O$

$$\frac{100}{140} \qquad \frac{2 \times 74}{L_3} \qquad \frac{2 \times 100}{S_3} \qquad \frac{58}{S_4}$$

$$L_3 = 207 \text{ mg/L} \qquad S_3 = 280 \text{ mg/L}$$

$$S_4 = 81 \text{ mg/L}$$

4. $MgSO_4 + Ca(OH)_2 \rightarrow Mg(OH)_2 + CaSO_4$

$$\frac{100}{60} \qquad \frac{74}{L_4} \qquad \frac{58}{S_5}$$

$$L_4 = 44 \text{ mg/L} \qquad S_5 = 35 \text{ mg/L}$$

5. $CaSO_4 + Na_2CO_3 \rightarrow CaCO_3 + Na_2SO_4$

$$\frac{100}{60} = \frac{106}{A_1} \quad \frac{100}{A_6}$$

$A_1 = 64$ mg/L $\quad S_6 = 60$ mg/L $\quad A_6 = 60$ mg/L

To decrease Mg^{++} from 33 mg/L to 12 mg/L as $CaCO_3$

$$\begin{array}{l} 62 \text{ mg/L of } OH^- \text{ alk as } CaCO_3 \text{ required} \\ -12 \text{ mg/L of } OH^- \text{ alk as } CaCO_3 \text{ to remain} \\ \hline 50 \text{ mg/L of } OH^- \text{ alk must be added} \end{array}$$

We will add the OH^- as $Ca(OH)_2$.

$$50 \text{ mg/L of } OH^- \text{ alk as } CaCO_3 \times \frac{37}{50} \text{ (equiv. wt. } Ca(OH)_2) =$$

37 mg/L of excess lime must be added
Expressed as lime, $L_5 = 37$ mg/L

But we have added Ca^{++} and must remove it!
Consider changing Ca^{++} to $CaCO_3$

$$Ca(OH)_2 + Na_2CO_3 \rightarrow CaCO_3 + 2NaOH$$

Set up proportion:

$$\frac{74}{37} = \frac{106}{A_2}$$

$$74A_2 = (37)(106)$$

$$A_2 = \frac{(37)(53)}{37}$$

$A_2 = 53$ mg/L of Na_2CO_3 must be added
To lower the Ca as $CaCO_3$ from 35 to 12 mg/L

$$\begin{array}{l} 59 \text{ mg/L of } CO_3 = \text{alk as } CaCO_3 \text{ required} \\ -12 \text{ mg/L of } CO_3 = \text{alk as } CaCO_3 \text{ to remain} \\ \hline 47 \text{ mg/L of } CO_3^- \text{ alk as } CaCO_3 \text{ must be added} \end{array}$$

To do this, we will add the carbonate alkalinity using sodium carbonate (Na_2CO_3).

47 mg/L of Na_2CO_3 as $CaCO_3$ must be converted to the amount of Na_2CO_3 expressed as Na_2CO_3.

$$47 \text{ mg/L} \times \frac{53}{50} = 49.8 \text{ mg/L} \quad \text{use 50 mg/L}$$

$$A_3 = 50 \text{ mg/L}$$

$$\text{Lime required} = L_1 + L_2 + L_3 + L_4 + L_5$$

$$18.5 + 148 + 207 + 44 + 37 = 454.5 \text{ mg/L} = 15{,}162 \text{ lb/day of 100\% lime}$$

Chemicals:

$$\text{Soda ash required} = 64 + 53 + 50 = 167 \text{ mg/L} = 5{,}571 \text{ lb/day of 100\% soda ash}$$

How much sludge? Add “S”s

REFERENCES

Powell, S. T., *Water Conditioning for Industry*, McGraw-Hill, New York, 1954.
Standard Methods for the Examination of Water and Wastewater, 18th ed., APHA, AWWA and WEF, 1992.

CHAPTER 3

Fundamental Principles and Concepts of Ion Exchange

This chapter provides a basic understanding of the ion exchange process, ion exchangers, and the terminology associated with the process and its design.

You will notice that the rigorous theory associated with the ion exchange process is absent. We recommend Helfferich's *Ion Exchange* (1962) and Weber's *Physiochemical Processes for Water Quality Control* (1972) to provide a strong theoretical background.

INTRODUCTION

In order to produce water or treat wastewater so that a very specific chemical composition is achieved, it is not usually feasible to use chemical precipitation. This is true even if several streams that are treated differently from one another are blended. The only readily usable means to achieve the desired effluent composition is ion exchange, where the replacement of one ion for another can be accomplished in a predictable manner. Most uses for ion exchange are where the primary objective is to remove and concentrate unwanted ions in the feed water to the exchanger. Because ion exchange is a stable and easily predictable exchange of unwanted for wanted ions, it can be used as a chemical feed mechanism as well as a removal process. Applications of specific ion exchange in the process industries are numerous, for example, recovering ionic metals, separating wanted from unwanted ionic constituents, and feeding medications into a wound or incision after surgery, to name just a few. A scan of *Chemical Abstracts* shows many more uses for ion exchange. These are all in addition to the normal uses such as hardness removal, deionization as an alternative to distillation, dealkalization, and heavy metal removal from wastewaters.

BASICS

Ion exchange describes a myriad of terms, such as the ion exchange process, the ion exchange phenomenon, and the unit process of ion exchange. It is referred to as both an adsorption process and a sorption process. However, ion exchange is most often defined as the reversible exchange of ions between a solid and a liquid in which there is no substantial change in the structure of the solid. Used to soften water by exchanging calcium and magnesium ions for sodium ions, ion exchange is also widely used to remove metals from industrial effluents. Ion exchange cannot effectively remove dissolved organics such as pesticides and polychlorinated biphenyls (PCBs) because these materials do not ionize. Dissolved organics can be removed effectively by polymeric adsorbants, which resemble ion exchange resins in that they too can be regenerated and reused.

The basic component of any ion exchange system is the insoluble solid, usually referred to as the ion exchanger or ion exchange resin. In most water and wastewater applications, the liquid is water and the solid is any number of natural or synthetic ion exchangers. A variety of natural and synthetic ion exchangers exist and are described in detail in this chapter.

Batch/Columnar Operation

Ion exchange processing can be accomplished by either a batch method or a column method. The batch method involves mixing the resin and solution in a batch tank, allowing the exchange to come to equilibrium, and then separating the resin from solution. The degree to which the exchange takes place is limited by the preference the resin exhibits for the ion in solution. Consequently, the utilization of the resin's exchange capacity is limited unless the selectivity for the ion in solution is far greater than for the exchangeable ion attached to the resin. Because batch-wise regeneration of the resin is chemically inefficient, batch processing by ion exchange has limited potential application and will not be discussed further.

Basic Columnar Operation

The primary function of most fixed-bed (columnar) ion exchange operations is the removal of an ionic species from solution in exchange for another ionic species for the purpose of concentrating a desirable material into a small volume or removing an undesirable constituent. The latter may be accomplished by exchange with an unobjectionable ionic species, for example, the softening of water using a cation exchanger in the sodium cycle. In most cases, the removal of all ions or particular ionic species must be as complete as possible and the ion exchange resin must have the capacity to treat a large volume of liquid per unit volume of exchanger.

In Chapter 5, we will see that ion exchange resins in fixed-bed operations are primarily designed for removing soluble ionic species and, therefore, cannot be expected to adsorb macro or colloidal ionic species except at the surface of the particle. Because of their ionic nature and particle size, ion exchange beds are very effective filters; however, they are not normally used for this purpose. If they are used and the filter load is heavy, one cannot expect them to function efficiently as ion exchangers.

Figure 3-1 shows a typical column operation used to soften a hard water containing calcium magnesium and sodium salts. Water with the indicated constituents is passed through an ion exchange column containing an ion exchanger or resin that has been regenerated such that the exchange sites are in the sodium form. The chemical constituents in the treated water are shown leaving the ion exchange column. Figure 3-2 shows typical hardware.

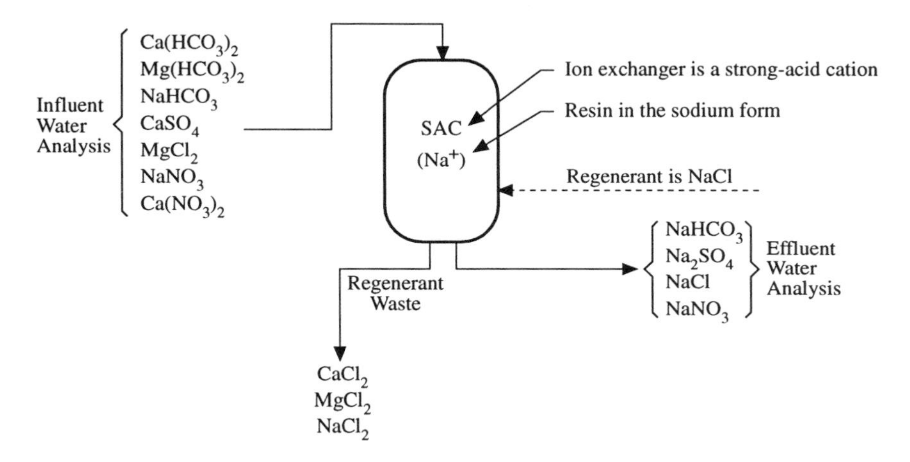

Figure 3-1 Typical columnar softening.

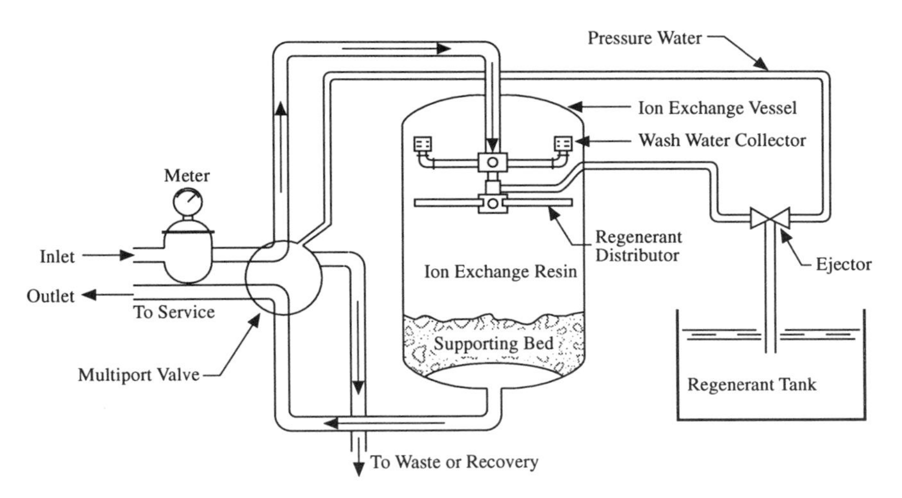

Figure 3-2 Typical ion exchange unit.

NATURAL ION EXCHANGERS

Many materials found in nature are able to capture certain ions and retain them in an exchangeable state. This is known as the exchange capacity of the material.

Most natural ion exchange materials are crystalline aluminosilicates with cation exchange properties (i.e., they remove positively charged cations from solution), although certain aluminosilicates can also act as anion exchangers. Included in the cation exchangers are the zeolites, analcity, chabazite, hormotome, hevlandite, and natrolite. Apatite and hydroxylapatite are two aluminosilicates used to remove anions in full-scale operations.

Zeolites are relatively soft minerals and are not very abrasion resistant. Their frameworks are less open and more rigid than those of most other ion exchangers. They, therefore, swell very little, and the counter ions in their pores are not very mobile.

Many coals are natural ion exchangers. They contain carboxylic and possibly other weak-acid groups and can thus be used as cation exchangers. Most of these materials, however, swell excessively, are easily decomposed by alkali, and tend to peptize. They must, therefore, be "stabilized" before use. Soft and hard lignitic coals have been stabilized by treatment with solutions of copper, chromium, or aluminum salts. Moreover, most lignitic and bituminous coals and anthracites can be converted into strong-acid cation exchangers by sulfonation with fuming sulfuric acid. Sulfonic acid groups are introduced and additional carboxylic acid groups are formed by oxidation.

A number of other natural materials exhibit ion exchange properties. Alumina, alginic acid, collodin, and keratin are a few typical examples of this group. A still larger number of materials can be transformed into ion exchangers by chemical treatment. From many soluble substances that carry ionogenic groups, insoluble ion-exchanger gels can be obtained simply by cross-linking with agents such as formaldehyde. Pectins and carrageen are typical examples. Conversely, from many insoluble substances ion exchangers can be produced by incorporation of fixed ionic groups. The most common procedures are sulfonation and phosphorylation of materials such as olive pits, nut shells, spent ground coffee, tar, wood, paper, cotton, lignins, and tannins.

Inorganic ion exchangers are used in both industrial and municipal water and wastewater treatment applications. The first applications, such as the use of aluminosilicates for softening, have been improved by the use of synthetic resins. Greensand (glauconite) is still used today in connection with iron and manganese removal. Activated alumina is used to remove trace inorganics such as fluoride, arsenic, selenium, and phosphate. Clinoptilolite is used to remove ammonia from water and wastewater.

One inorganic material in particular has been shown to be an effective ion exchange medium for the removal of heavy metals. Etzel and Keramida used the clay mineral vermiculite — a layered silicate consisting of two tetrahedral sheets composed of aluminosilicates and a central trioctahedral layer — to remove copper, zinc, and nickel from synthetic and actual wastewaters. Work is in progress to further develop the concept.

SYNTHETIC ION EXCHANGERS

The majority of ion exchange resins are made by the copolymerization of styrene and divinylbenzene (DVB). The styrene molecules provide the basic matrix of the resin, and the DVB is used to cross-link the polymers to allow for the general insolubility and toughness of the resin. The degree of cross-linking in the resin's three-dimensional array is important because it determines the internal pore structure, which in turn affects the internal movement of exchanging ions.

Visualize the synthetic resin shown in Figure 3-3 as a "whiffle ball" — a skeleton-like structure having many exchange sites. The skeleton, insoluble in water, is electrically charged, holding ions of opposite charge at the exchange sites.

Synthetic resins are available in bead form and range in size from 20 mesh (0.84 mm in diameter) to 325 mesh (0.044 mm in diameter). Most ion exchange applications in water and wastewater are accomplished with resins in the 20 to 50 mesh size range.

Recent developments in polymer chemistry have resulted in the production of macroporous (or macroreticular) resins that have a discrete pore structure. These resins are more resistant to thermal and osmotic shock as well as oxidation. The more porous resins are also more resistant to organic fouling than gel-type resins.

Table 3-1 shows that synthetic exchangers are generally divided into four major classifications depending on the resin's functional group (but two types of weak base). The functional group determines whether cations or anions are exchanged and whether the resin is a strong or weak electrolyte.

Synthetic Cation Exchangers

Cation exchangers are available with numerous fixed ionic groups exhibiting a range of different properties and acid strengths. The most common are strong-acid resins (SAR) with sulfonic groups ($-SO_3-$) and weak-acid resins (WAR) with carboxylic acid groups and phosphonic acid groups.

In general, strong-acid cation exchangers substitute one ion for another depending on the resin's selectivity, and as such operate at all pH values. Their capacity is less than stoichiometric, however, and they must be regenerated more frequently than weak-acid resins, which exhibit much higher capacities and regenerate almost stoichiometrically. Weak-acid resins operate only over a limited pH range.

Strong-Acid Cation Exchange Resins

All major strongly acidic cation exchange resins involved in water and wastewater treatment applications have a chemical matrix consisting of styrene and DVB. The functional groups are sulfonic acid radicals. These resins differ mainly in DVB content and gelular or macroreticular structure.

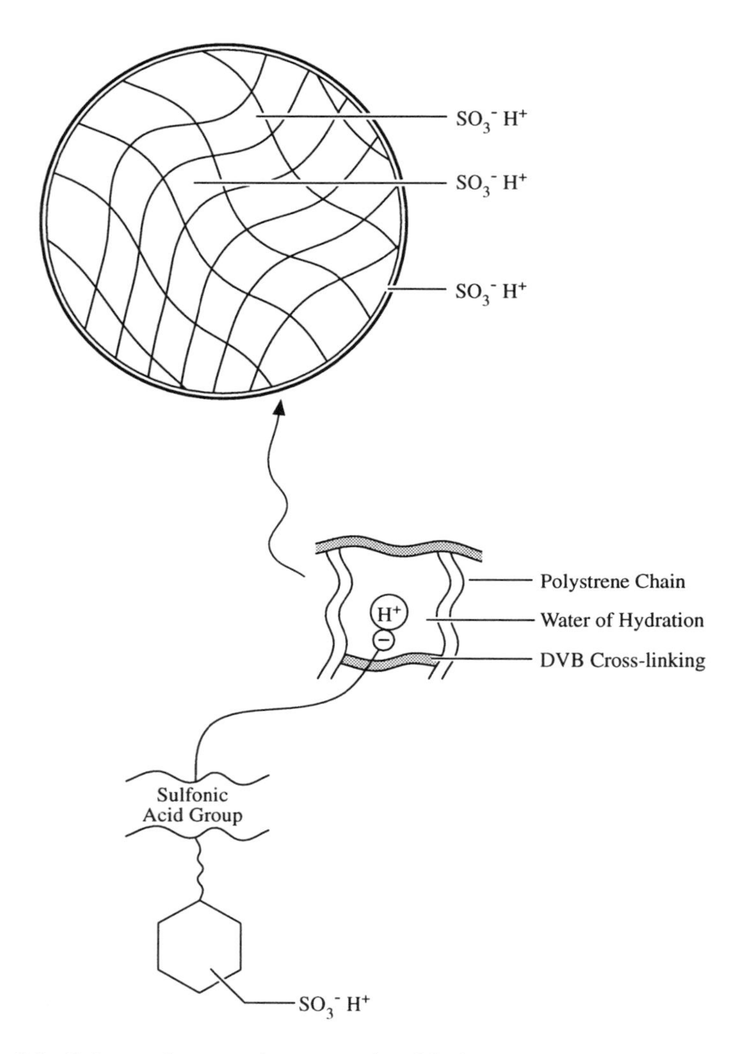

Figure 3-3 Cation exchange resin — strongly acidic (hydrated).

Strong-acid resins are so named because their chemical behavior resembles that of a strong acid. A strong-acid cation resin can convert a neutral salt into its corresponding acid if operated in the acid form. This ability is known as salt-splitting and distinguishes strong-acid resins from weak-acid resins, which cannot salt-split. The equation for salt-splitting is

$$NaCl + (resin - H^+) \rightarrow HCl + (resin - Na^+)$$

Strong-acid exchangers can be operated in any form, but are typically operated in the hydrogen cycle, where the resin is regenerated with a strong acid such as HCl or H_2SO_4, or in the sodium cycle where the resin is regenerated with NaCl.

Table 3-1 Classification of the Major Ion Exchange Resins

Type	Active group	Typical configuration
	Cation Exchange Resins	
Strong acid	Sulfonic acid	(benzene ring)–SO_3H
Weak acid	Carboxylic acid	—CH_2CHCH_2— with COOH on CH
	Anion Exchange Resins	
Strong base	Quaternary ammonium	(benzene ring)–$CH_2N(CH_3)_3Cl$
Weak base	Secondary amine	(benzene ring)–CH_2NHR
Weak base	Tertiary amine (aromatic matrix)	(benzene ring)–CH_2NR_2

Author's note: Probably the most important thing you will get from this book is that a resin (in this case a strong-acid resin) can be regenerated or put into any cycle that you wish. Ion exchangers can be operated in the K^+ cycle, the Ca^{++} cycle, the Mg^{++} cycle — any cycle that will allow the ion exchanger to work for you.

A strong-acid cation exchanger in the hydrogen cycle will remove nearly all major raw water cations and is the first step in demineralizing water. The equation for removing calcium sulfate from a raw water with an ion exchanger in the hydrogen cycle is

$$CaSO_4 + 2(R\text{–}H^+) \rightleftharpoons 2(R^-)\text{–}Ca^{++} + H_2SO_4$$

Figure 3-4 shows the same water treated by the sodium cycle exchanger in Figure 3-1, treated by a strong-acid cation resin in the hydrogen cycle. The resin in the bed is regenerated by either HCl or H_2SO_4. There will be further discussion of regeneration later in the chapter. *In general, a strong acid can be regenerated using any monovalent, divalent, or trivalent acid or salt.*

The strong-acid exchangers require excess strong-acid regenerant (typical regeneration efficiency varies from 25–45% in concurrent regeneration), and they permit low leakage. In addition, they have rapid exchange rates, are stable, and may last 20 years or more with little loss of capacity. They exhibit less than 7%

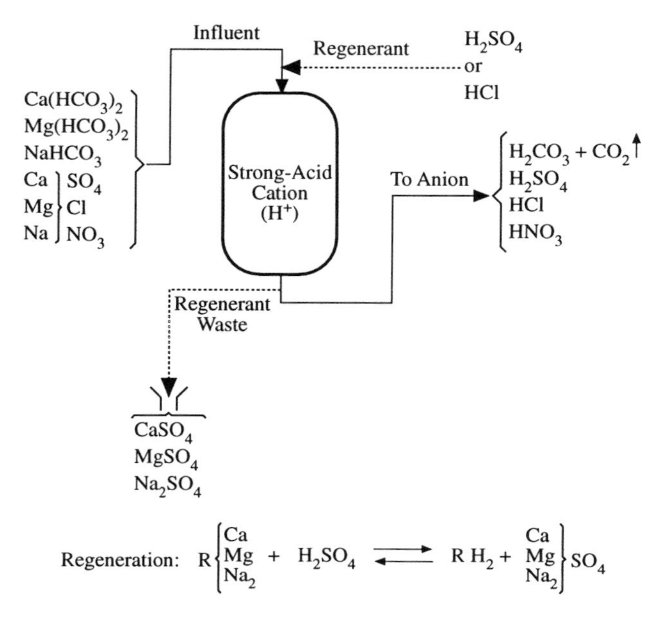

$$\text{Regeneration: } R\begin{Bmatrix}Ca\\Mg\\Na_2\end{Bmatrix} + H_2SO_4 \rightleftharpoons RH_2 + \begin{Bmatrix}Ca\\Mg\\Na_2\end{Bmatrix}SO_4$$

Figure 3-4 Hydrogen-cycle strong-acid cation exchange.

swelling going from Na^+ to H^+ form and are useful for softening and demineralization (removal of all cations with little leakage).

Generally, in most domestic water softening, one would use a strongly acidic cation exchange resin with an 8% DVB cross-linking. However, if a condition called oxidative decross-linking should occur, it would be best to use a resin with greater resistance to oxidative attack. Amberlite IR-122 resin, for example, with its higher DVB content is typical. In very severe decross-linking problems, a macroreticular type is recommended, where up to 20% DVB can be used as the cross-linker.

Weak-Acid Cation Exchange Resins

Referring to Table 3-1, carboxylic acid is the functional group associated with weak-acid cation exchangers. As such weak-acid cation resins are not highly dissociated and do not exchange their H^+ as readily as strong resins. But because they exhibit a higher affinity for hydrogen ions than strong-acid resins, weak-acid resins show higher regeneration efficiencies. They are usually regenerated with strong acids such as hydrochloric or sulfuric acids. Weak-acid resins do not require the same concentration driving force required to convert strong-acid resins to the hydrogen form. The carboxylic functional groups have a high affinity for H^+ and will utilize up to 90% of the acid (HCl or H_2SO_4) regenerant, even with low acid concentrations. This is in contrast to strong-acid resin regeneration, where a large excess of regenerant (of which 60–75% goes unutilized) is required to create the concentration driving force.

Because of their high affinity for the hydrogen ion, weak-acid resins can be used only at pH's above 4 or 5, as shown in Figure 3-5.

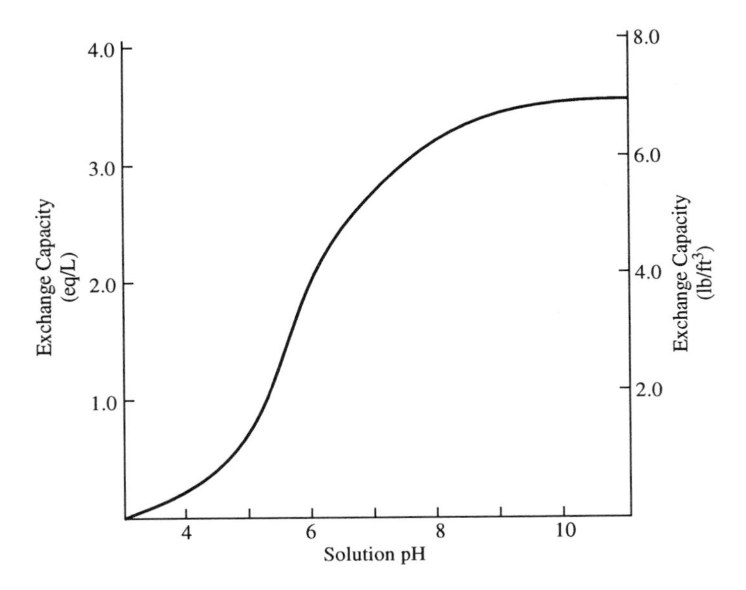

Figure 3-5 Capacity of weak-acid cation resin as a function of solution pH.

Weak-acid exchangers differ from strong-acid resins in that weak-acid resins require the presence of some alkaline species to react with the more tightly held hydrogen ions of the resin, for example,

$$Ca(HCO_3)_2 + 2(R\text{–}H^+) \rightleftharpoons (2R^-)\text{–}Ca^{++} + 2(H_2CO_3)$$

The exchange is, in effect, a neutralization with the alkalinity (HCO_3^-) neutralizing the H^+ of the resin. Weak-acid resins will split alkaline salts but not nonalkaline salts (e.g., $NaHCO_3$ but not NaCl or Na_2SO_4).

Weak-acid cation resins are used to remove the cations associated with high alkalinity, i.e., $CO_3^=$, OH^-, and HCO_3^-, and low in dissolved CO_2 and sodium. Weak-acid resins are used primarily for achieving simultaneous softening and dealkalization. They are sometimes used in conjunction with a strong-acid polishing resin, which allows for economic operation in terms of regenerant requirements, but also produces a treated water of quality comparable to the use of just a strong-acid resin.

Synthetic Anion Exchangers

Anion exchangers were developed almost exclusively as synthetic resins; organic exchangers were among the earliest ion exchange resins produced. The first patents issued for anion exchangers were for resins having weak-base amino groups. Later, resins with strong-base quaternary ammonium groups were prepared.

Anion exchangers are available with numerous fixed ionic groups. The most common are strong-base resins (SBR) with quaternary ammonium groups

($-CH_2N(CH_3)_3Cl$) and weak-base resins (WBR) with tertiary amines in an aromatic or aliphatic matrix.

In general, strong-base anion exchangers operate at all pH values, but their capacity is less than stoichiometric, and they must be regenerated more frequently than weak-base resins, which exhibit much higher capacities and regenerate almost stoichiometrically. Weak-base resins operate only over a limited pH range.

Strong-Base Anion Exchange Resins

Strong-base resins are available. In Type I resins, the functional groups consist of three methyl groups

$$-N^{+}- \quad -N^{+}(CH_3)_3-$$

In Type II resins, an ethanol group replaces one of the methyl groups. Type I resins exhibit greater stability. Type II resins exhibit slightly greater regeneration efficiency and capacity.

Strong-base exchangers are so named because they have the ability to split strong or weak salts. This ability distinguishes them from weak-base resins, which cannot salt-split.

The reactions with sulfate and chloride and a strong-base anion exchanger in the hydroxyl form are

$$\left.\begin{matrix} SO_4^{=} \\ 2Cl^- \end{matrix}\right\} + 2R-(OH)^- \rightleftharpoons 2R-\begin{pmatrix} SO_4^{=} \\ 2Cl^- \end{pmatrix} + 2(OH)^-$$

Figure 3-6 shows the same water treated by the hydrogen cycle exchanger in Figure 3-4. The resin in this case is regenerated by sodium hydroxide.

In general, when placed in the hydroxide form, strong-base exchangers require an excess of regenerant (with typical efficiencies varying 18–33%). Typically, high-quality sodium hydroxide is used as the regenerant, but again, the regenerant used depends on the desired form in which the resin is to be used. A problem with strong-base resins is that they tend to irreversibly sorb humic acid substances, losing capacity. Activated carbon or a weak-base organic trap is typically used to prevent fouling.

Type I exchangers are typically used for maximum silica removal. They are more difficult to regenerate and swell more (from Cl^- to OH^- form) than Type II. The principal use of Type I is to make the highest quality water. When they are loaded with silica, they must be regenerated with warm NaOH. (See resin manufacturer's literature.)

Weak-Base Anion Exchangers

The functional groups associated with weak-base anion exchangers are secondary and tertiary amines and are often based on phenol-formaldehyde or epoxy

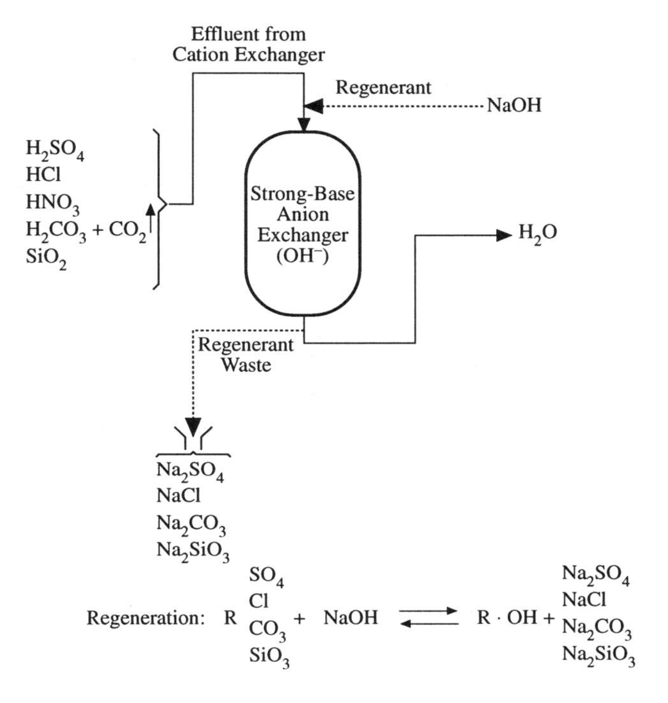

$$\text{Regeneration: } R \begin{matrix} SO_4 \\ Cl \\ CO_3 \\ SiO_3 \end{matrix} + NaOH \rightleftharpoons R \cdot OH + \begin{matrix} Na_2SO_4 \\ NaCl \\ Na_2CO_3 \\ Na_2SiO_3 \end{matrix}$$

Figure 3-6 Strong-base ion exchange reactions.

matrices instead of polystyrene-DVB. They do not remove anions above a pH of 6 (see Figure 3-7). They regenerate with a nearly stoichiometric amount of base (with the regeneration efficiency possibly exceeding 90%) and are resistant to organic fouling. In addition, they swell about 12% going from the OH^- to salt form, they do not remove CO_2 or silica, and they have capacities about twice as great as that of strong-base exchangers. They are useful following strong-acid exchangers to save cost of regenerant chemicals, as organic "traps" to protect strong-base exchangers, and to remove color.

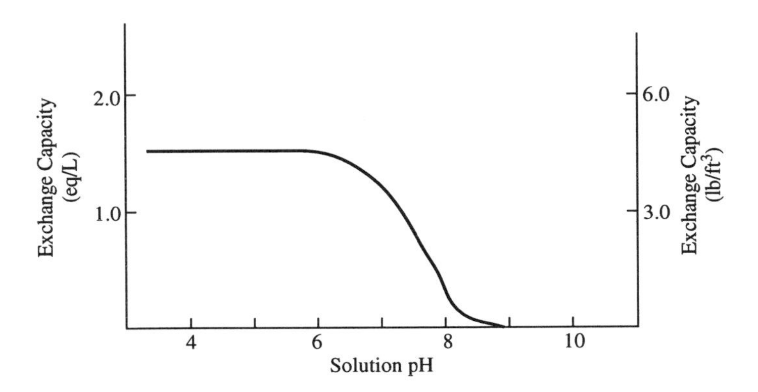

Figure 3-7 Capacity of weak-base anion resin as a function of solution pH.

The weak-base anion exchange resins behave much like their weak-acid counterparts and will not remove carbon dioxide, i.e., weakly ionized acids such as carbonic and silicic. The weak-base resins remove free mineral acidity (FMA) such as HCl or H_2SO_4, i.e., anions associated with the hydrogen ion and that are strong acid formers such sulfate, chloride, or nitrate. For this reason, weak-base resins are often called "acid adsorbers." Any ions associated with cations other than hydrogen are not removed.

The weak-base resin does not have a hydroxide ion form as does the strong-base resin. Consequently, regeneration need only neutralize the absorbed acid, not provide hydroxide ions, and less expensive weakly basic reagents such as ammonia or sodium carbonate may be used, along with sodium hydroxide.

Once again, the regeneration efficiencies of these resins are much greater than those for strong-base resins. Weak-base exchangers are used in conjunction with strong-base resins in demineralizing systems to reduce regenerant costs and to attract organics that might otherwise foul the strong-base resins. Where silica removal is not critical, weak-base resins may be used by themselves in conjunction with an air stripper to remove CO_2 (see Chapter 5).

The major advantage of weak-base anion exchange resins is that they can be regenerated with stoichiometric amounts of regenerant and are, therefore, much more efficient. They also have a much higher capacity for the removal of chlorides, sulfates, and nitrates. There are three typical weakly basic anion exchange resins: styrene-DVB, acrylic-DVB, and epoxy. Again, gel or macroreticular varieties of resins based on these raw materials are available and used. For treatment of waters that do not present organic fouling problems, the gel type of weakly basic resin is used.

For waters containing organic contaminants (humic and fulvic acids), macroreticular weakly basic anion resins are preferred. Since their development in the late 1960s, acrylic resins are beginning to be used more, and are found to be more effective from an economic standpoint. Figure 3-8 shows typical weak-acid and weak-base ion exchange applications.

SELECTIVITY

The most important concept presented in this book is as follows: *The success of any ion exchange system is dependent on the cycle in which the resin is operated*, i.e., *operate the ion exchange resin in the proper cycle*. Do not limit yourself to the hydrogen cycle or the sodium cycle, the chloride cycle or the hydroxyl cycle. Make the ion exchange resin work for you.

Each ion exchange resin has its own order of exchange preference. In general, trivalent is preferred over divalent, which is preferred over monovalent. A bumping order based on valence exists. A bumping order based on atomic number ions of the same valence exists.

Order of Selectivity (Relative Affinity)

In general, the higher the valence of an ion, the higher its ability to be removed by a resin. For example, for a resin in the hydrogen form, if a divalent cation came along, it would be taken preferentially to the monovalent cation (H^+) on

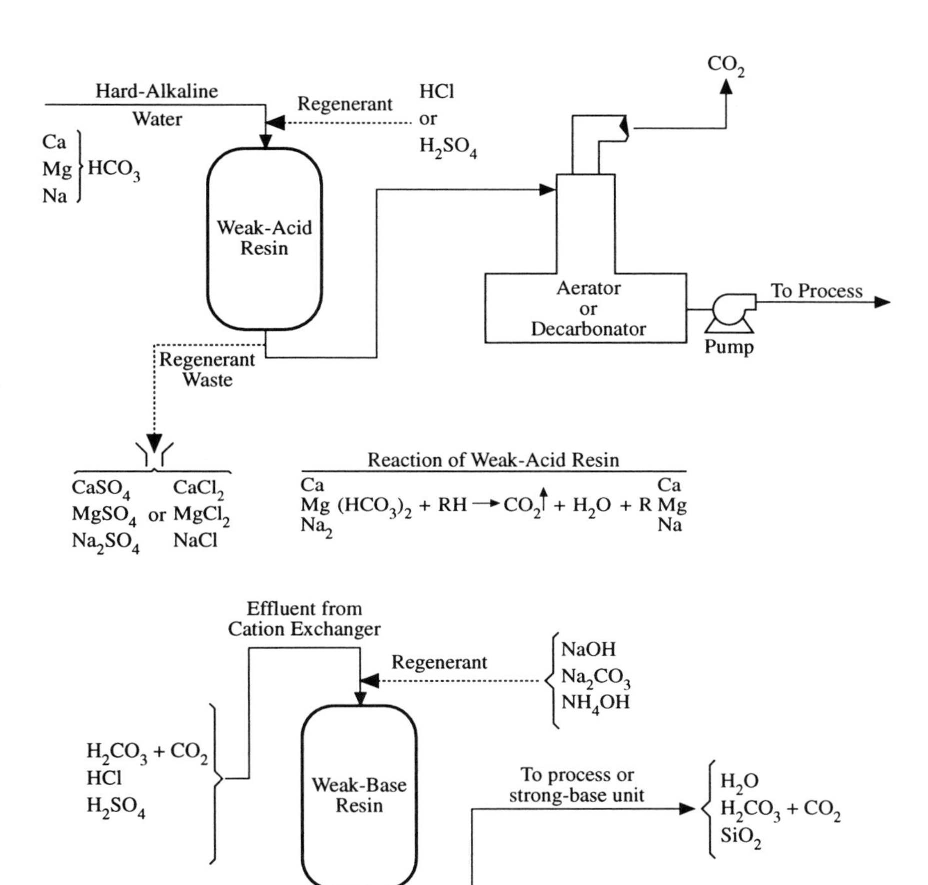

Figure 3-8 Weak-acid and weak-base ion exchange reactions.

the resin. If a trivalent cation came along, it would preferentially displace the divalent cation (in nearly all circumstances).

Now imagine the ion exchange material is in the hydrogen form. If a sodium ion is introduced, the sodium ion would displace the hydrogen ion, i.e., the sodium ion would stay on the resin and the hydrogen ion would be carried out. If we then introduce a calcium ion, the calcium ion would displace two sodium ions and the calcium ion would stay on the resin. If we then introduce aluminum ions,

Table 3-2 Order of Selectivity — Typical Strong-Acid Resin

Ion	Valence
Barium	+2
Strontium	+2
Calcium	+2
Magnesium	+2
Beryllium	+2
Silver	+2
Thallium	+1
Cesium	+1
Rubidium	+1
Ammonia	+1
Potassium	+1
Sodium	+1
Hydrogen	+1
Lithium	+1

Note: This order will vary with different resins, but usually the order shifts only by one or two. Trivalent replaces divalent, which replaces monovalent ions.

Source: Dr. Etzel's class notes.

Table 3-3 Order of Selectivity — Typical Strong-Base Resin

Ion	Valence
Thiocyanate	–3
Iodide	–2
Nitrate	–1
Bromide	–1
Cyanide	–1
Bisulfite	–1
Nitrite	–1
Chloride	–1
Bicarbonate	–1
Acetate	–1
Hydroxide	–1
Fluoride	–1

Source: Dr. Etzel's class notes.

two aluminum ions would displace six calcium ions. Table 3-2 shows the order of selectivity of a typical strong-acid resin. Table 3-3 shows the order of selectivity of a typical strong-base resin.

In the design of ion exchange systems, one must be careful about the cycle in which one chooses to operate the resin, because this will determine whether one gets the job done.

Changing the degree of cross-linking does not change the order of selectivity but does change relative affinities.

As shown in Table 3-4, the higher the number, the easier for that ion to displace ions lower than it. For example, to remove Ag from a wastewater stream

Table 3-4 Selectivity Scale for Selected Cations on Dowex Cation Resin at 4, 8, and 16% DVB Cross-Linking

	X4[a]	X8[a]	X16[a]
	Monovalent		
Li	1.00	1.00	1.00
H	1.32	1.27	1.47
Na	1.58	1.98	2.37
NH_3OH	1.90	2.25	3.28
NH_4	1.90	2.55	3.34
K	2.27	2.90	4.50
Rb	2.46	3.16	4.62
Cs	2.67	3.25	4.66
Ag	4.73	8.51	22.9
Tl	6.71	12.4	28.5
	Divalent		
UO_2	2.36	2.45	3.34
Mg	2.95	3.29	3.51
Zn	3.13	3.47	3.78
Co	3.23	3.74	3.81
Cu	3.29	3.85	4.46
Cd	3.37	3.88	4.95
Be	3.43	3.99	6.23
Ni	3.45	3.93	4.06
Mn	3.42	4.09	4.91
Ca	4.15	5.16	7.27
Sr	4.70	6.51	10.1
Pb	6.56	9.91	18.0
Ba	7.47	11.5	20.8
	Trivalent		
Cr	6.60	7.60	10.5
Ce	7.50	10.6	17.0
La	7.60	10.7	17.0

[a] Cross-linking with divinylbenzene at 4, 8, and 16%.
Note: Lithium is arbitrarily set at 1.00.
Source: Dr. Etzel's class notes.

use Dowex cation resin with 16% cross-linking. If there is any sodium in water, with a selectivity of 2.37, Ag will easily displace it and the Ag is monovalent. If you are trying to remove a specific ion, you want the highest spread you can get, and if you can get different levels of cross-linking to do that, even better. A minimum selectivity delta of two is required for most exchanges.

This order varies with different resins (depending on the manufacturer), but usually the order is shifted by only a few elements. Trivalent replacing divalent replacing monovalent is always the same.

Table 3-4 also shows the relative affinity of selected cation resins as a function of valence and cross-linking with divinylbenzene at 4, 8, and 16%. Lithium is assigned a value of 1.00. Table 3-5 presents the relative selectivity of selected

Table 3-5 Order of Selectivity for Dowex Strong-Base Anion Resin

Dichlorophenate	53
Salicylate	28
Phenate	8.7
Iodide	7.3
Bisulfate	6.1
Nitrate	3.3
Bromide	2.3
Nitrite	1.3
Cyanide	1.3
Bisulfite	1.3
Bromate	1.01
Chloride	1.00
Hydroxide	0.65
Bicarbonate	0.53
Dihydrogen phosphate	0.34
Formate	0.22
Acetate	0.18
Fluoride	0.13
Amino acetate	0.10

Note: Chloride is arbitrarily set at 1.00.

anions for Dowex strong-base anion resin. Chloride is assigned a value of 1.00. Contact the resin manufacturer you have chosen for the relative exchange order of a particular resin.

Use the following rules as a guide in understanding the relative selectivity patterns of exchange.

1. At low aqueous concentrations and ordinary temperatures, the extent of exchange or the exchange potential increases with increasing valence of the exchanging ion.

   $Na^{+} < Ca^{++} < Al^{+++}$

2. At low aqueous concentrations, ordinary temperatures, and constant valence, the exchange potential increases with increasing atomic number.

   $Li < Na < K$; $Mg < Ca < Sr < Ba$; $F < Cl < Br < 1$

3. At high concentrations, the differences in exchange potentials of ions of different valence diminish and, in some cases, the ion of lower valence has the higher exchange potential.

   Na^{+} vs. Ca^{++}

4. The exchange potentials of various ions may be approximated from their activity coefficient — the higher the activity coefficient the greater the potential.
5. The exchange potential of the hydrogen ion or the hydroxyl ion depends on the strength of the acid or base formed between the functional group and the hydrogen or hydroxyl ion. The stronger the acid or base the lower the potential.

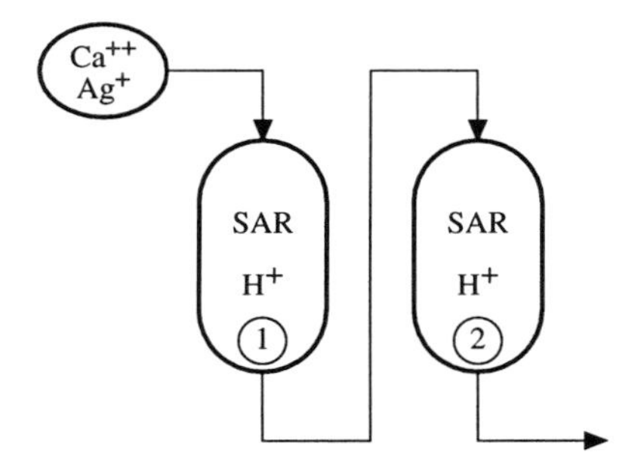

Figure 3-9 Silver separation diagram.

Selectivity Determination (see Chapter 5)

Select the resin. Put the resin in the lithium cycle. Run sodium ions through and monitor how much sodium stays on the resin and how much runs through. Do this until sodium-in = sodium-out. Analyze the resin to determine how much sodium is on the resin. Repeat for every ion. Compare the percent of material retained as a percent of lithium (e.g., retains twice as much lithium, i.e., 2.0).

Example: How to use a resin's selectivity to separate silver from calcium (see Figure 3-9).

Column 1 removes both calcium and silver, but eventually the calcium displaces the silver. If we take column 2 off-line before any calcium breaks through column 1, we will have only silver on the resin. We have separated silver from calcium.

Remember:

We can regenerate a resin with any compound that will dissolve in water and ionize.
We can put a resin in any form we want.

RESIN STRUCTURE

Think of a synthetic ion exchange resin as a network of hydrocarbon radicals to which soluble ionic functional groups are attached, with the hydrocarbon network usually formed from the copolymerization of styrene and DVB. Overall, the insolubility and toughness is given to the resin by cross-linking between the hydrocarbon molecules, forming a three-dimensional matrix. The internal pore structure of the resin is determined by the extent or degree of cross-linking. Cross-linking adds stability to the styrene molecule and increases the resin's ability to resist acid/base attack. Because exchanging ions must be free to move into and out of the resin for exchange to occur, the degree of cross-linking should not be so great as to restrict this free movement of ions, although in some cases cross-

Table 3-6 Physical Characteristics of Amberlite IR-120 Plus

Physical form — Hard, attrition resistant, light yellow, 16–50 mesh (U.S. Standard Screens), fully hydrated spherical particles
Shipping weight — 53 lb/ft^3 (848 g/L) (sodium form)
— 50 lb/ft^3 (800 g/L) (hydrogen form)
Moisture content — 45%
Effective size — 0.50 mm
Uniformity coefficient — 1.8 maximum
Density — 53 lb/ft^3 (848 g/L)
Void volume — 35 to 40%

Sodium or hydrogen cycle

pH	1.0 to 14.0
Maximum temperature	250°F (121°C)
Minimum bed depth	24 in. (0.61 m)
Backwash flow rate	See detailed information in text
Service flow rate	2 gpm/ft^3 (16.0 L/L/hr)

Source: Rohm and Haas.

linking can exclude ions larger than a given size from the resin. Table 3-6 shows the physical characteristics of Amberlite IR-120 Plus, a strong-acid resin.

As you have seen and will see, resin characteristics such as selectivity and capacity are almost exclusively determined by the nature of the ionic groups attached to the framework of the resin.

CAPACITY

On a quantitative basis, the capability of a resin to exchange ions is defined as its exchange capacity, with total capacity being defined as the number of ionic sites per unit weight or volume of resin. Because it is impractical to use all available sites during exchange, the net number of sites used in a specific volume of resin in a given cycle is defined as the operating capacity. From a design point of view, resin capacity, that is, the number or quantity of counter ions (the ions in solution that we want to remove) that can be taken up by the exchanger (resin), is probably the most important property of ion exchange resin. The exchange capacity of a resin comes from the substituted functional groups in the resin matrix.

At this time, it is important to distinguish between the total capacity of a resin and its operating capacity. Total capacity is a measure of the total number of counter ions that can be exchanged. For strong-acid cation exchange resins, one sulfonate group, on the average, can be attached to each benzene ring in the matrix. Hence, we can determine the *dry-weight capacity* of this resin, which would be expressed in milliequivalents per gram of dry resin (meq/g). The dry-weight capacity is simply a measure of the extent of functional group substitution in the exchanger and is therefore a constant for each specific resin. For sulfonated styrene-DVB resins, the dry-weight capacity is 5.0 ± 0.5 meq/g (Anderson, 1979).

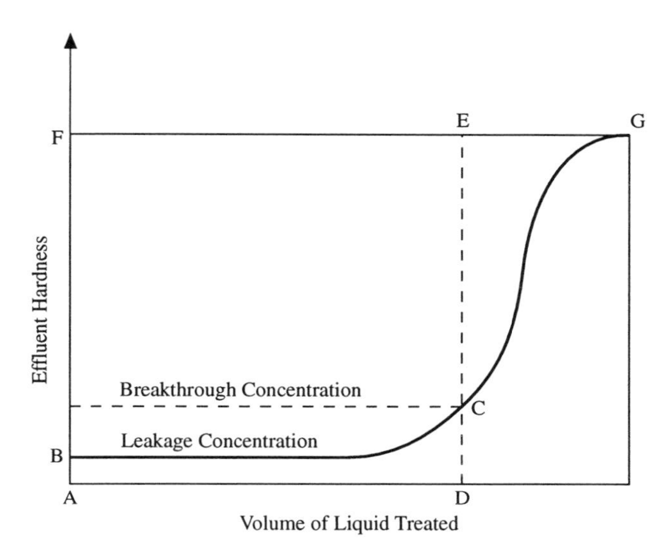

Figure 3-10 Typical breakthrough curve for an ion exchange process.

For strong-base anion exchangers, more or less than one functional group can be attached to each benzene ring. Hence, the dry-weight capacity is more variable than with strong-acid resins and can range from 2 to 5 meq/g.

Operating capacity is the exchange capacity of a resin operated in a columnar mode under defined operating conditions and a predetermined breakthrough level. Breakthrough occurs when a predetermined concentration of a target ion or ions (hardness, for example) occurs in the effluent of an ion exchange system. Breakthrough is chosen arbitrarily. Figure 3-10 shows a typical breakthrough curve for an ion exchange system treating hardness.

Leakage is the result of acid (or whatever regenerant is used) being formed during the exchange process. The regenerant formed at the top of the bed may regenerate an easily exchanged site (on the resin surface) lower down in the bed, allowing the exchanged material to escape into the effluent. A statistical probability exists of an ion entering the bed and missing all of the exchange sites, but this is minimized by having the bed always at least 30" deep. A plot of exchange capacity vs. the amount of regenerant acid used shown on the curve in Figure 3-11 is approximate. Note that on the left side (near origin) there is a large change in capacity for very little acid. This change in slope indicates that there must be some sites that are very easy to exchange and some that are very difficult to exchange.

When the bed is exhausted the acid gradient goes back to zero because it can no longer produce acid. The maximum acid gradient occurs in the bed just before leakage occurs. The leakage problem is most prevalent in hydrogen cycle systems because any acid can cause regeneration.

Note: When using the high-capacity part of the resin, there is a higher probability of leakage. To obtain more capacity, we must exchange more sodium onto

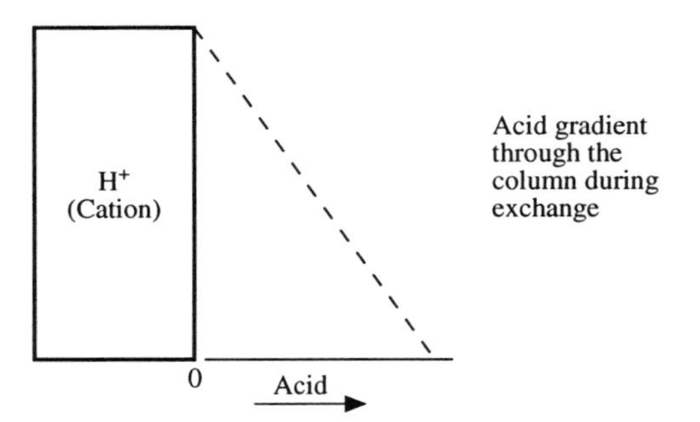

Figure 3-11 Acid gradient.

the resin, including the less efficient sites. With more sites available for exchange, less chance exists of internal regeneration. At the low end of the resin, expect more leakage.

Leakage and breakthrough can be controlled by using two or more beds in series. When leakage begins to occur, regenerate the first bed and put it on-line after the second bed as a polishing unit (see Figure 3-9). Leakage can also be nearly eliminated by using a mixed-bed (hydrogen and hydroxyl form) unit so that the acid (regenerant) produced is destroyed as soon as it is formed.

The operating capacity of a resin is usually expressed in a weight per volume capacity, that is, meq/ml — milliequivalents of ions exchanged per milliliter of resin. Other units for expressing capacity are lb-equivalents/ft^3, grams of $CaCO_3$ per liter, and, the most used, kilograins/ft^3.

The natural exchange material greensand has a capacity of 2–5 Kgr/ft^3 of sand. Zeolite (silica gels) typically exhibit capacities of 3–12 Kgr/ft^3. Synthetic resin capacities range from 3 to 50 Kgr/ft^3, with a usable range of 15–25 Kgr/ft^3.

A few numbers to remember

$$7000 \text{ grains} = 7 \text{ kilograin (Kgr)} = 1 \text{ pound}$$

$$17.12 \text{ mg/L} = \frac{1 \text{ grain}}{\text{gallon}}$$

PARTICLE SIZE

A few words about particle size are in order. Ion exchange resins in spherical shapes are available commercially in particle diameter sizes of 0.04–1.0 mm. In the United States, the particle sizes are listed according to standard screen sizes or "mesh" values. Table 3-7 shows a comparison of mesh sizes and metric sizes. The most common size ranges used in large-scale applications are 20–50 and 50–100 mesh.

Manufacturers typically provide information about three parameters related to particle size: particle size range, effective size (ES), and uniformity coefficient

Table 3-7 Ion Exchange Particle Size Availability

U.S. mesh	Diameter (mm)
16–20	1.2–0.85
20–50	0.85–0.30
50–100	0.30–0.15
100–200	0.15–0.08
200–400	0.08–0.04

(UC). The size range gives the maximum and minimum sizes of particles in the batch, the ES is a screen size that passes 10% (by weight) of the total quantity while 90% is retained, and the UC is the ratio of the mesh size (in millimeters) that passes 60% of the quantity to the ES. Typical UCs are in the range of 1.4–1.6; however, it is possible to obtain batches with smaller uniformity coefficients if required by kinetic or hydraulic restrictions. In Figure 3-12, the effective size is A. The UC is B/A.

Particle size has two major influences on ion exchange applications. First, the kinetics of exchange are such that the rate of exchange is proportional to either the inverse of the particle diameter or the inverse of the square of the particle diameter. Second, particle size has a great effect on the hydraulics of column design. Smaller particle sizes increase pressure drops through the bed, requiring a higher head to push the water through the resin beads and subjecting the beads to situations that could cause breakage. In 50% of all ion exchange applications, the design is based on hydraulic limitations of the resin beads and the vessel rather than on ion exchange chemistry.

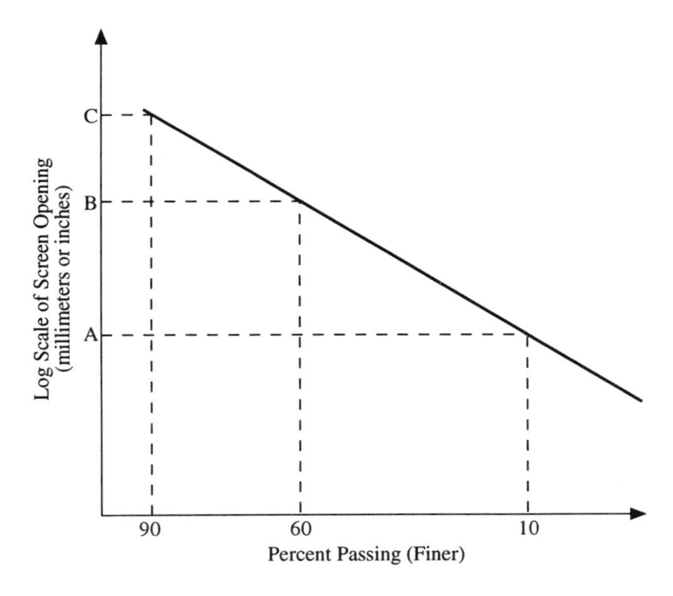

Figure 3-12 Resin sieve analysis.

PROCESS DESCRIPTION

The columnar ion exchange process is best illustrated by considering the basic steps of the complete cycle of an ion exchange operation. The operation of any ion exchange process involves four steps: backwash, regeneration or brining, rinse (displacement, fast and final), and service.

Backwash

Backwash is carried out with a new installation and after each "run." Water is introduced through the bottom of the ion exchange column at a rate sufficient to expand the bed by 50–75%. The primary purpose of the backwash step is to remove any silt, dirt, iron, or other insoluble matter that has accumulated during the exchange operation. Also, any clumps of resin formed because of tight packing are dispersed, and the resin beads are reoriented. Most of the problems associated with ion exchange operation occur because of insufficient backwash.

Figure 3-13 provides hydraulic expansion data for Amberlite IR-120. It shows bed expansion in percent as a function of temperature and flow rate in (gpm/ft^2). Typical flow rates are upflow at 5–7 gpm/ft^2 of exchanger cross-sectional area.

Regeneration

Although the columnar ion exchange process consists of the basic steps of backwash, regeneration, rinsing, and service, the actual process is regeneration with the appropriate backwash and rinsing steps and the service cycle. A more in-depth discussion of regeneration is in order.

Regeneration usually uses 5–15% of the water originally treated. The regeneration actually consists of three steps:

1. Backwashing — wash out any dirt in bed
2. Regeneration — actual chemical addition
3. Rinsing — rinse out regeneration chemicals

The regeneration itself uses only 5% of the water treated and thus all of the ions removed will be concentrated in about 5% of the water → 20 times as concentrated.

Recall that the ion exchange resin looks very much like a whiffle ball with exchange sites on the inside and the outside. If insoluble compounds such as calcium sulfate are formed upon regeneration it can block or plug the internal exchange sites.

Regeneration or brining follows backwash and is the displacement of the ions held on the resin sites. These ions were removed from the process feed water during the service cycle. If the column contains a strong-acid resin regenerated with an acid, for example, hydrogen ions (H^+) are exchanged onto the resin in place of those released. This resin is now in the hydrogen form. If the column is a strong-base resin and is regenerated with sodium hydroxide, hydroxyl ions

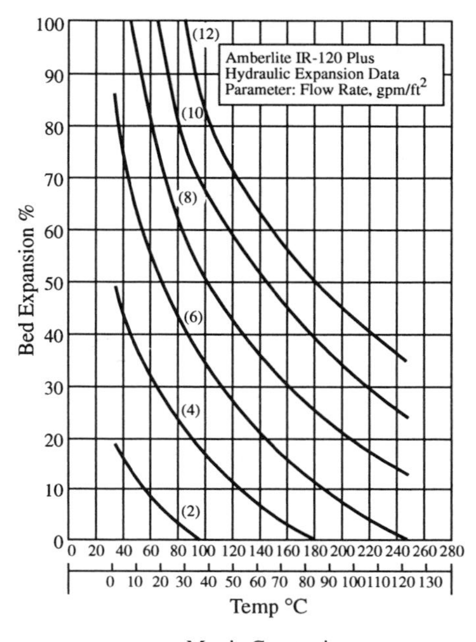

Figure 3-13 Hydraulic backwash data. (From Rohm and Haas, Amberlite® IR-120 Plus Specification Sheet, January 1982. With permission.)

(OH^-) are exchanged on the resin. It is then in the hydroxyl form. The regeneration step can be accomplished in either a downflow or upflow mode. When softening water or using the column in the sodium cycle, the regeneration step is often referred to as "brining." In most cases, concentrated sodium chloride is passed through the bed to drive the calcium and magnesium (softening) or other metals (industrial uses) off of the resin. Again, brining can be accomplished in either an upflow or downflow mode. Table 3-8 provides typical regeneration data for regenerating a Rohm and Haas strong-acid ion exchange resin.

Note that resin specifications will always show you that there is a greater resin capacity when hydrochloric acid is used than when sulfuric acid is used. As an example, you might get 22 Kgr/ft^3 using hydrochloric acid and only 18 Kgr/ft^3 using sulfuric acid. The reason for this is that any calcium exchanged onto the resin comes off during regeneration as $CaSO_4$, which has a very limited solubility of only about 2200 mg/L. Thus if there is more than 110 mg/L Ca in the raw water there will probably be a solubility problem with H_2SO_4 regeneration. $CaCl_2$ is almost infinitely soluble, so there is no problem with HCl.

When using a sodium cycle system to soften water containing iron you will experience a slow progressive decrease in capacity over time. This is because iron is being exchanged onto internal sites and becoming oxidized to the ferric form (insoluble), thus blocking exchange sites. This effect can be minimized by more frequent regeneration, so that the iron has less time to oxidize and block

Table 3-8 Resin Specifications for Amberlite IR-120 Plus — A Strong-Acid Cation Exchanger

The recommended regeneration conditions for hydrogen cycle operation of Amberlite IR-120 Plus are listed below:

REGENERANT CONCENTRATION[1] — 10% HCl or 1 to 5% H_2SO_4

REGENERANT FLOW RATE — 0.5 to 1.0 gpm/ft^3 (4.0 to 8.0 L/L/hr)

RINSE FLOW RATE — Initially same as regenerant flow rate, then can be increased to 1.5 gpm/ft^3 (12.0 L/L/hr)

RINSE WATER REQUIREMENTS — 25 to 75 gal/ft^3 (3.4 to 10.1 L/L)

REGENERATION — The tables below show the relationship between capacity and levels of sulfuric and hydrochloric acid for regeneration. Sulfuric acid concentration used after NaCl exhaustion was 10%. After $CaCl_2$ exhaustion, regeneration using 2% sulfuric acid was employed to avoid calcium sulfate precipitation. A 10% solution of hydrochloric acid was used in both NaCl and $CaCl_2$ exhaustion studies.

(1) *Caution*: Nitric acid and other strong oxidizing agents can cause explosive type reactions when mixed with organic materials, such as, ion exchange resins. Before using strong oxidizing agents in contact with ion exchange resins, consult sources knowledgeable in the handling of these materials.

Acid Regeneration

Exhausting solution (ppm as $CaCO_3$)	Regeneration level (lb of 66° Be' H_2SO_4/ft^3 of resin)	g acid/L resin	Capacity $\left(\frac{\text{Kgr.}^{(2)} \text{ as } CaCO_3}{\text{ft}^3 \text{ resin}}\right)$	g $CaCO_3$/L
500 ppm	5.0	80	19.0	43.5
NaCl	10.0	160	25.0	57.3
500 ppm	5.0	80	12.5	28.6
$CaCl_2$	10.0	160	17.0	38.9

Exhausting solution (ppm as $CaCO_3$)	Regeneration level (lb 30% HCl/ft^3 of resin)	g acid/L resin	Capacity $\left(\frac{\text{Kgr.}^{(2)} \text{ as } CaCO_3}{\text{ft}^3 \text{ resin}}\right)$	g $CaCO_3$/L
	5	80	11.0	25.2
500 ppm	15	240	22.5	51.5
$CaCl_2$	25	400	27.5	63.0

(2) Kgr. = kilograins

Amberlite IR-120 Plus will provide excellent performance in both cold sodium cycle softeners and hot process systems. The recommended regeneration conditions for sodium cycle operation are listed below:

REGENERANT CONCENTRATION — 10% NaCl

REGENERANT FLOW RATE — 1 gpm/ft^3 (8.0 L/L/hr)

RINSE FLOW RATE — 1 gpm/ft^3 (8.0 L/L/hr) initially, then 1.5 gpm/ft^3 (12.0 L/L/hr)

RINSE WATER REQUIREMENTS — 25 to 75 gal/ft^3 (3.4 to 10.1 L/L)

REGENERATION — The relationship between regeneration level and capacity is summarized in the table below. Data were obtained using 500 ppm (as $CaCO_3$) calcium chloride solution. Capacities have been adjusted downward to typify performance of material meeting *minimum* production specifications.

Table 3-8 (continued)

Regeneration level (lbs $NaCl/ft^3$ resin)	g NaCl/L resin	Capacity (Kgr. as $CaCO_3/ft^3$ resin)	g $CaCO_3$/L resin	Regeneration efficiency (lbs NaCl/Kgr. removed)	g NaCl/ g $CaCO_3$ removed
5.0	80	17.8	40.8	0.28	1.96
15.0	240	29.3	67.1	0.51	3.57
25.0	400	34.0	77.9	0.74	5.13

Note: Acidic and basic regenerant solutions are corrosive and should be handled in a manner that will prevent eye and skin contact. In addition, the hazards of alcohols and other organic solvents should be recognized and steps taken to control exposure.

the resin. Even with frequent regeneration there will still be some loss of capacity. The resin manufacturer will recommend that once or twice per year you regenerate with sodium chloride mixed with sodium bisulfite. The sodium bisulfite is such a strong reducing agent that it will reduce the insoluble ferric iron to soluble ferrous form so that you can clean up the bed. Usually 1 or 2 oz. of sodium bisulfite per ft^3 of resin is sufficient.

Rinse

Rinsing follows regeneration. The purpose is to remove excess regenerant prior to putting the unit into service. This is usually downflow. Softening operations involve a displacement rinse, a fast rinse, and a final rinse. After brining, the resin bed is full of brine. During the displacement rinse, the brine (concentrated NaCl) is slowly forced through the resin bed. The objective is to obtain optimum contact. The displacement rinse is always done at the same rate and direction as brining. Raw water is used for all rinsing, except in certain dionization operations. A high flow rate is used to remove any residual brine from the resin bed. The fast rinse is usually in the same direction as the displacement rinse (see Table 3-8).

Service

Service is the operational mode of the system. This step follows rinsing. Process water, for example, plating rinse water, is passed through the ion exchanger to remove cations or anions. The end of the service run is detected by a sharp increase in the cation or anion level of the effluent. In most industrial applications, an acceptable or predetermined level of cations or anions in the effluent is set. When this level is reached, "breakthrough" has occurred. The unit is then taken off-line and is backwashed, regenerated, and rinsed before being put back into service. (see Table 3-5 and Figure 3-14.)

Pretreatment of process water prior to ion exchange is crucial to prevent fouling or damage to the resin. At a minimum, the water must be filtered to remove suspended solids. Cartridge filters work best because they do not require backwashing (another side stream requiring disposal). Organic substances in the

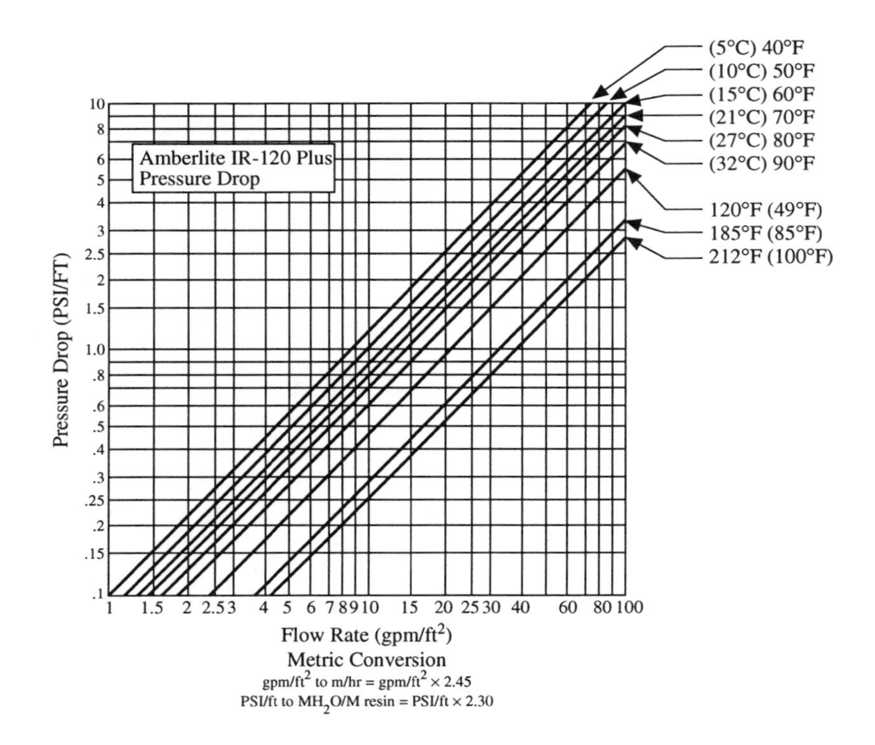

Figure 3-14 Pressure drop versus flow rate. (From Rohm and Haas, Amberlite® IR-20 Plus Specification Sheet, January 1982. With permission.)

process water can foul most resins and should be removed by pretreatment with activated carbon. Fouling occurs when organics in the water affix themselves to the resin by occupying the exchange sites. This phenomenon decreases the capacity of the resin. Strong oxidizing agents (e.g., chlorine) degrade most resins. These agents must be removed or destroyed by a pretreatment step.

EXAMPLE PROBLEMS, CHAPTER 3

REAL-WORLD REGENERATION

EXAMPLE PROBLEM:

A chemical plant produces ammonium nitrate from nitric acid and ammonia, both of which they make. Their raw water contains only 40 mg/L of total dissolved solids (TDS) expressed as $CaCO_3$. The wastewater from the plant contains 200 mg/L of NH_3-N and 250 mg/L of NO_3-N in addition to the original TDS. If they use 8 lb/ft^3 of HNO_3 and 8 lb/ft^3 of NH_4OH to get 22 and 11 Kgr/ft^3 of capacity, respectively, as $CaCO_3$ on their resins when the regenerants have a concentration of 8% by weight, what volume of regeneration wastes do they have per day? What is the composition of the regeneration wastes in bar graph form? The cost of reusing the deionized water produced in waste treatment is too high to warrant its recovery. (Assume complete removal of all ions in the raw wastewater.)

SOLUTION

The NH_3-N was there as NH_4OH-N

The NO_3^--N was there as HNO_3-N

$$NH_4OH + HNO_3 \rightarrow NH_4NO_3$$

$$\frac{14}{200\ \text{mg/L}} = \frac{14}{x} = \frac{14}{y}$$

x = 200 mg/L HNO_3-N used to produce 200 mg/L of NH_4NO_3-N

∴ 50 mg/L HNO_3-N unreacted

$$\frac{200\ \text{mg/L}\ NH_4NO_3\text{-N}}{14} = \frac{x}{50}$$

x = 714 mg/L NH_4NO_3-N as $CaCO_3$

$$\frac{50\ \text{mg/L}\ HNO_3\text{-N}}{14} = \frac{y}{50}$$

y = 178.6 mg/L HNO_3-N as $CaCO_3$

Cations to be removed = 40 + 714 = 754 mg/L as $CaCO_3$

$$754\ \text{mg/L} \div 17.12\ \text{mg/L/gr/gal} = 44\ \text{gr/gal as}\ CaCO_3$$

Anions to be removed = 40 + 714 + 178.6 = 932.6 mg/L as $CaCO_3$

$$932.6\ \text{mg/L} \div 17.12\ \text{mg/L/gr/gal} = 54.5\ \text{gr/gal as}\ CaCO_3$$

Volume of anion resin

$$54.5\ \text{gr/gal} \times 10^5\ \text{gal} \times \frac{\text{Kgr}}{10^3\ \text{gr}} = 5450\ \text{Kgr/day as}\ CaCO_3$$

$$5450 \text{ Kgr/day} \div 11 \text{ Kgr/ft}^3 = 495.5 \text{ ft}^3 \text{ of resin}$$

Volume of cation resin

$$44 \text{ gr/gal} \times \left(10^5 \text{ gal} + 495.5 \text{ ft}^3 \times 70 \text{ gal/ft}^3\right) \times \frac{\text{Kgr}}{10^3 \text{ gr}} = 5926 \text{ Kgr/day as CaCO}_3$$

$$5926 \text{ Kgr/day} \div 22 \text{ Kgr/ft}^3 = 269.4 \text{ ft}^3 \text{ of resin}$$

Design of cation units

Assume three units with a 24-hr regeneration cycle

$$134{,}682 \text{ gal/day} \div 1440 \text{ min/day} = 93.5 \text{ gal/min}$$

With two units on-line flow rate is

$$93.5 \text{ gal/min} \div 2 = 46.8 \text{ gal/min/unit}$$

At 5 gal/min/ft^2

$$46.8 \text{ gal/min/unit} \div 5 \text{ gal/min/ft}^2 = 9.4 \text{ ft}^2\text{/unit}$$

A 42"-diam. unit will have an area of 9.62 ft^2

$$269.4 \text{ ft}^3 \div 3 \text{ units} \times 9.62 \text{ ft}^2\text{/unit} = 9.33 \text{ ft deep (ok)}$$

Design of anion units

Assume three units with a 24-hr regeneration cycle
With two units on line flow rate is

$$100{,}000 \text{ gal/day} \div 1440 \text{ min/day} \times 2 \text{ units} = 34.7 \text{ gal/min/unit}$$

At 2.5 gal/min/ft^2

$$34.7 \text{ gal/min/unit} \div 2.5 \text{ gal/min/ft}^2 = 13.88 \text{ ft}^2$$

A 54"-diam. unit will have an area of 15.9 ft^2

$$495.5 \text{ ft}^3 \div 3 \text{ units} \times 15.9 \text{ ft}^2\text{/unit} = 10.4 \text{ ft depth (ok)}$$

Regeneration needs

$$\text{Cation}\quad 269.4\ \text{ft}^3 \times 8\ \text{lb HNO}_3/\text{ft}^3 = 2155.2\ \text{lb/day HNO}_3$$

$$\text{Anion}\quad 495.5\ \text{ft}^3 \times 8\ \text{lb NH}_4\text{OH/ft}^3 = 3964\ \text{lb/day NH}_4\text{OH}$$

$$2155.2\ \text{lb/day HNO}_3 \times \frac{50}{63} = 1710.5\ \text{lb/day HNO}_3\ \text{as CaCO}_3$$

$$3964\ \text{lb/day NH}_4\text{OH} \times \frac{50}{35} = 5662.9\ \text{lb/day NH}_4\text{OH as CaCO}_3$$

$$1710.5\ \text{lb/day HNO}_3\ \text{as CaCO}_3 \times 7\ \text{Kgr/lb} = 11{,}973.5\ \text{Kgr/day as CaCO}_3$$

$$5662.9\ \text{lb/day NH}_4\text{OH as CaCO}_3 \times 7\ \text{Kgr/lb} = 39{,}640.3\ \text{Kgr/day as CaCO}_3$$

$$\begin{array}{rl} 11{,}973.5 & \text{Kgr/day HNO}_3\ \text{as CaCO}_3\ \text{used} \\ -\ \underline{5926} & \text{Kgr/day cations as CaCO}_3\ \text{exchanged} \\ 6047.5 & \text{Kgr/day HNO}_3\ \text{as CaCO}_3\ \text{not reacted} \end{array}$$

$$\begin{array}{rl} 39{,}640.3 & \text{Kgr/day NH}_4\text{OH as CaCO}_3\ \text{used} \\ -\ \underline{5450} & \text{Kgr/day anions as CaCO}_3\ \text{exchanged} \\ 34{,}190.3 & \text{Kgr/day NH}_4\text{OH as CaCO}_3\ \text{not reacted} \end{array}$$

$$\begin{array}{ccccccc} \text{NH}_4\text{OH} & + & \text{HNO}_3 & \rightarrow & \text{NH}_4\text{NO}_3 & + & \text{H}_2\text{O} \\ 6047.5 & & 6047.5 & & 6047.5 & & 6047.5 \end{array}$$

If all of the above are expressed as $CaCO_3$ then

$$34{,}190.3 - 6047.5 = 28{,}142.8\ \text{Kgr of NH}_4\text{OH as CaCO}_3\ \text{unreacted in regen. waste}$$

$$40\ \text{mg/L TDS as CaCO}_3 \div 17.12\ \text{mg/L/gr/gal} = 2.34\ \text{gr/gal TDS as CaCO}_3$$

$$2.34\ \text{gr/gal} \times 134{,}682\ \text{gal/day} \times \frac{\text{Kgr}}{10^3\ \text{gr}} = 315\ \text{Kgr TDS as CaCO}_3\text{/day}$$

$$\text{(Cation)}\quad 5926 - 315 = 5611\ \text{Kgr NH}_4\text{NO}_3\ \text{as CaCO}_3\ \text{in regen. waste}$$

$$\text{(Anion)}\quad 5450 - 315 = 5135\ \text{Kgr NH}_4\text{NO}_3\ \text{as CaCO}_3\ \text{in regen. waste}$$

Thus

5611 Kgr NH_4NO_3 as $CaCO_3$ produced from regen. of cation units

5135 Kgr NH_4NO_3 as $CaCO_3$ produced from regen. of anion units

6047.5 Kgr NH_4NO_3 as $CaCO_3$ produced from reaction of excess NH_4OH as $CaCO_3$ and HNO_3 as $CaCO_3$ in regen. waste

28,142.8 Kgr NH_4OH as $CaCO_3$ from excess NH_4OH used in regen.

315 Kgr XNO_3 as $CaCO_3$ from regen. of cation units

315 Kgr NH_4X as $CaCO_3$ from regen. of anion units

6047.5 Kgr H_2O as $CaCO_3$ from reaction of NH_4OH and HNO_3 in regen. wastes

Volume of regen. wastes

$$2155.2 \text{ lb } HNO_3 \div 0.08 = 26,940 \text{ lb/day acid wastes}$$

$$3964 \text{ lb } NH_4OH \div 0.08 = 49,550 \text{ lb/day basic wastes}$$

$$6047.5 \text{ Kgr } H_2O \text{ as } CaCO_3 \div 7 = 864 \times \frac{9}{50} = 155.5 \text{ lb/day } H_2O \text{ produced}$$

$$\text{Total} = 76,645.5 \text{ lb/day regen. waste}$$

$$76,645.5 \text{ lb/day} \div 8.34 \text{ lb/gal} = 9190 \text{ gal/day regen. waste}$$

$$16,793.5 \text{ Kgr } NH_4NO_3 \text{ as } CaCO_3 \div 9190 \text{ gal} = 1827.4 \text{ gr/gal as } CaCO_3$$

$$28,142.8 \text{ Kgr } NH_4OH \text{ as } CaCO_3 \div 9190 \text{ gal} = 3062.3 \text{ gr/gal as } CaCO_3$$

$$315 \text{ Kgr } XNO_3 \text{ as } CaCO_3 \div 9190 \text{ gal} = 34.3 \text{ gr/gal as } CaCO_3$$

$$315 \text{ Kgr } NH_4X \text{ as } CaCO_3 \div 9190 \text{ gal} = 34.3 \text{ gr/gal as } CaCO_3$$

0		4924	4958.3
NH_4^+			x^+
OH^-	NO_3^-		x^-
	3062.3	4924	4958.3

EXAMPLE PROBLEM: REGENERATION WASTE

Calculate the concentrations of cations and anions expressed as $CaCO_3$ in the regeneration wastewater for a sodium softener treating 2.0 MGD of raw water to a residual of 1.2 gr/gal. The raw water analyses and bar graph of the raw water are shown below. Draw a bar graph of the regeneration water. Assume the resin you have chosen is Amberlite® IR-120 Plus, with a regeneration efficiency of 0.51 lb NaCl per kilograin of hardness removed.

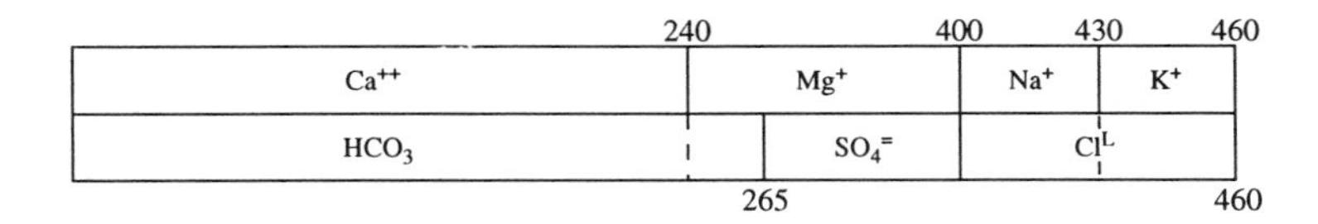

Raw Water Analyses

$$Ca(HCO_3)_2 = 240 \text{ mg/L as } CaCO_3$$

$$Mg(HCO_3)_2 = 25 \text{ mg/L as } CaCO_3$$

$$MgSO_4 = 115 \text{ mg/L as } CaCO_3$$

$$Na_2SO_4 = 20 \text{ mg/L as } CaCO_3$$

$$NaCl = 30 \text{ mg/L as } CaCO_3$$

$$KCl = 30 \text{ mg/L as } CaCO_3$$

$$NO_3 = 0 \text{ mg/L as } CaCO_3$$

Design

Hardness (Ca^{++} and Mg^{++}) = 380 mg/L as $CaCO_3$

$$380 \text{ mg/L} \times \frac{1 \text{ gr/gal}}{17.12 \text{ mg/L}} = 22.2 \text{ gr/gal as } CaCO_3$$

Since hardness removal via softening by ion exchange is essentially an all-or-nothing process, we must design the softener with some flow by passing the softener to obtain a finished water of 1.2 gr/gal.

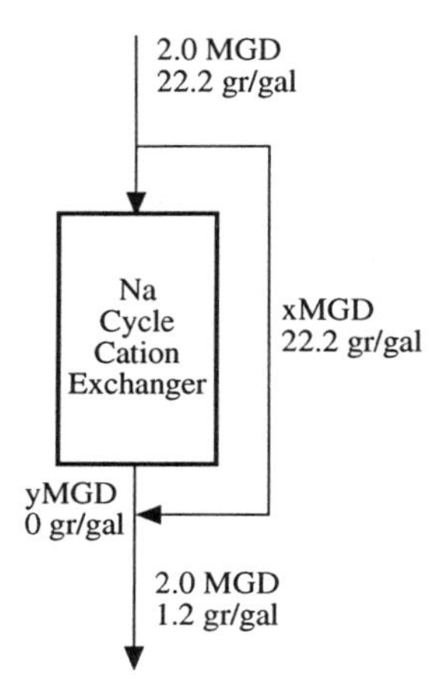

Assume the resin you have chosen is Amberlite IR-120 Plus, with a regeneration efficiency of 0.51 lb NaCl per kilograin of hardness removed:

$$\frac{(1.89 \times 10^6 \text{ gal/day})(22.2 \text{ gr/gal})(0.51 \text{ lb NaCl/Kgr})}{10^3 \text{ gr/Kgr}} = 21{,}398.6 \text{ lb/day}$$

NaCl required is 21,400 lb/day

Regeneration requires a volume of water equal to 6% of that produced between regenerations:

$$(0.06)(2 \times 10^6 \text{ gal}) = 120{,}000 \text{ gal} = 0.12 \text{ MGD of water for regeneration}$$

Calculate hardness removed each day:

$$(1.89 \text{ MGD})(380 \text{ mg/L})(8.34) = 5990 \text{ lb/day Ca and Mg as } CaCO_3 \text{ removed}$$

Calculate mg/L of hardness in regeneration wastewater:

$$5990 \text{ lb/day} = (x)(0.12 \text{ MGD})(8.34)$$

x = 5.985 mg/L of hardness as $CaCO_3$ in regeneration wastewater from unit

$$\text{Total Ca + Mg in regeneration wastewater} = 5985 \text{ mg/L} + 380 \text{ mg/L} = 6365 \text{ mg/L}$$

(380 mg/L is the water used for regeneration)

$$Ca = 6365 \times \frac{240}{380} = 4020 \text{ mg/L as } CaCO_3$$

$$Mg = 6365 \times \frac{140}{380} = 2345 \text{ mg/L as } CaCO_3$$

Calculate chloride in regeneration wastewater:

$$21{,}398.4 \text{ lb/day NaCl} \times \frac{50}{58.45} = 18.305 \text{ lb/day NaCl as } CaCO_3$$

$$18{,}305 \text{ lb/day} = (0.12 \text{ MGD})(8.34)(x)$$

$$x = 18{,}290 \text{ mg/L } Cl^- \text{ as } CaCO_3 \text{ in regen.}$$

$$\text{Total } Cl^- \text{ as } CaCO_3 \text{ in regen.} = 18{,}290 + 60 = 18{,}350 \text{ mg/L}$$

Calculate sodium in regeneration wastewater:

$$21{,}398.6 \text{ lb/day NaCl} \times \frac{50}{58.45} = 18{,}305 \text{ lb/day NaCl as } CaCO_3$$

$$-5990 \text{ lb/day exchanged on the resin}$$

$$= 12.315 \text{ lb/day NaCl as } CaCO_3 \text{ not exchanged on resin}$$

$$12{,}315 = (0.12 \text{ MGD})(8.34)(x)$$

$$x = 12{,}305 \text{ mg/L}$$

$$\text{Total } Na^+ = 12{,}305 + 50 = 12{,}355 \text{ mg/L as } CaCO_3 \text{ in regen. wastewater}$$

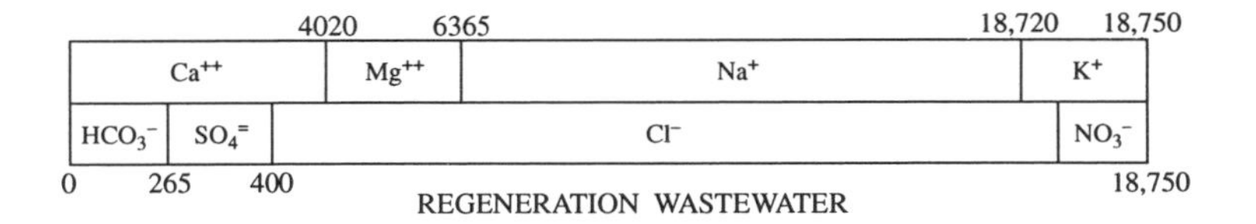

REGENERATION WASTEWATER

EXAMPLE PROBLEM

Starting with a typical midwestern groundwater, would you get more treatment capacity from a cation-anion deionization if you first treated water with sodium cycle softener or if you did not treat it at all? Explain.

SOLUTION

If the water is not softened, it will contain both divalent and monovalent cations, which can cause displacement of monovalent ions with divalent ions and thus leakage or premature breakthrough. Treatment with a softener would add only monovalent ions. Displacement will not take place and as such more water can be treated with the same volume of resin. So, the answer is *yes*.

EXAMPLE PROBLEM

Starting with water that contains the following:

$$H^+ = 100 \text{ mg/L as } CaCO_3$$

$$Al^{+++} = 60 \text{ mg/L as } CaCO_3$$

$$Na^+ = 40 \text{ mg/L as } CaCO_3$$

$$K^+ = 50 \text{ mg/L as } CaCO_3$$

$$SO_4^{=} = 140 \text{ mg/L as } CaCO_3$$

$$Cl^- = 110 \text{ mg/L as } CaCO_3$$

Produce 100,000 gal/day of water with a hardness of 33 mg/L as $CaCO_3$, 80 mg/L of hydroxide alkalinity as $CaCO_3$. No suspended solids may be in the finished water (filtration may not be used). You have three cation and two anion exchange units available and they can be operated in any reasonable cycle of exchange.

1. Show the set up of the system.
2. Indicate the cycle in which each unit must operate.
3. Calculate the flow through each unit.
4. Calculate the TDS in the finished water.

SOLUTION

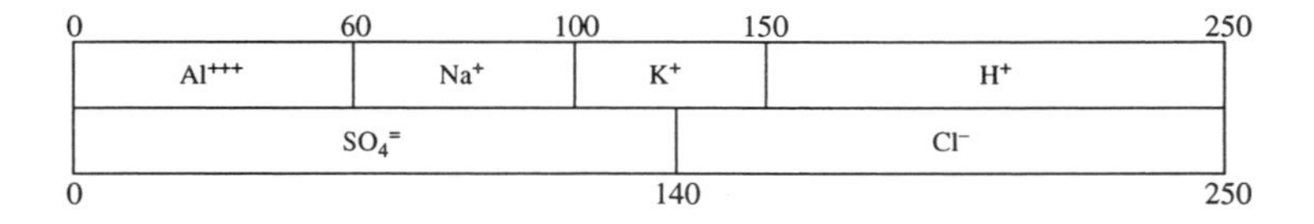

$$Al_2(SO_4)_3 = 60 \text{ mg/L as } CaCO_3$$
$$Na_2SO_4 = 40 \text{ mg/L as } CaCO_3$$
$$K_2SO_4 = 40 \text{ mg/L as } CaCO_3$$
$$KCl = 10 \text{ mg/L as } CaCO_3$$
$$HCl = 100 \text{ mg/L as } CaCO_3$$

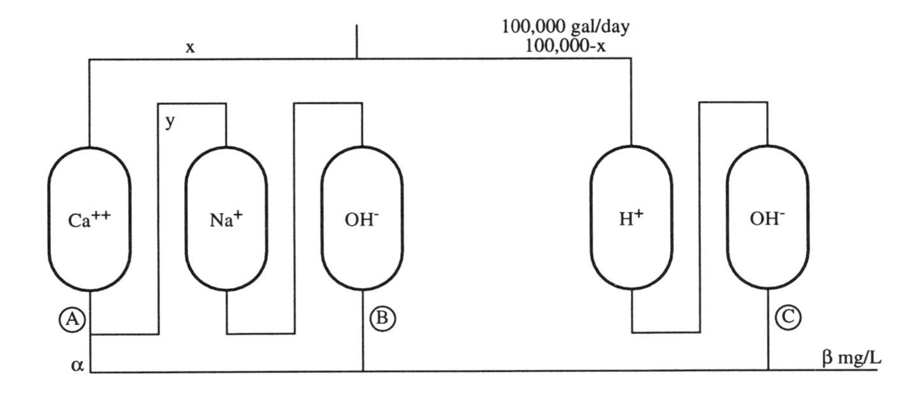

Ⓐ $HCl = 100 \text{ mg/L as } CaCO_3$

$CaSO_4 = 60 \text{ mg/L as } CaCO_3$

$K_2SO_4 = 40 \text{ mg/L as } CaCO_3$

$Na_2SO_4 = 40 \text{ mg/L as } CaCO_3$

$KCl = 10 \text{ mg/L as } CaCO_3$

Ⓑ $NaOH = 100 \text{ mg/L as } CaCO_3$

$KOH = 50 \text{ mg/L as } CaCO_3$

Ⓒ 0 TDS

Hardness

$$60 \text{ mg/L} \times \alpha \text{ gal} = 33 \text{ mg/L} \times 100,000 \text{ gal}$$

$$\alpha = 55,000 \text{ gal}$$

Alkalinity-acidity

$$-100 \text{ mg/L} \times 55,000 \text{ gal} + 150 \text{ mg/L} \times y = 100,000 \text{ gal} \times 80 \text{ mg/L}$$

$$y = 16,667 \text{ gal}$$

$$x = \alpha + y = 55,000 + 16,667 = 71,667 \text{ gal}$$

TDS

$$250 \text{ mg/L} \times 55,000 \text{ gal} + 150 \text{ mg/L} \times 16,667 \text{ gal} = \beta \times 100,000 \text{ gal}$$

$$\beta = 162.5 \text{ mg/L}$$

SHORT EXAMPLE: ION EXCHANGE

Aluminum has been linked with Alzheimer's disease. Consider a drinking water containing calcium, magnesium, sodium, and potassium with a trace (2 mg/L of Al^{+++}). In what cycle could a cation exchange be run to remove only the aluminum? With what would you regenerate the resin?

SOLUTION

$CaCl_2$, chloride cycle.

EXAMPLE PROBLEM

An ion exchange unit in the sodium cycle contains 15 ft^3 of resin and requires 10 lb NaCl ft^3 for regeneration. The volume of untreated water used for regeneration is 4% of the softened water produced. The analysis of the raw water is as follows: pH, 7.2; alk as $CaCO_3$, 240 mg/L; Ca^{++} as Ca^{++}, 57 mg/L; Mg^{++} as Mg^{++}, 48 mg/L; Na^+ as Na^+, 26.5 mg/L; $SO_4^=$ as $SO_4^=$, 154 mg/L. The capacity of the resin is 28 Kgr/ft^3 as $CaCO_3$.

1. Draw a bar graph showing the probable makeup of the chemical constituents in the raw water. Give a numerical value for each constituent.
2. Draw a bar graph showing the probable makeup of the chemical constituents in the treated water. Give a numerical value for each constituent.
3. What volume of treated water is produced per cycle?
4. Draw a bar graph showing the makeup of the probable chemical constituents in the regeneration wastes. Give a numerical value for each constituent.

SOLUTION

Ca^{++} 57 mg/L $\quad \frac{57}{20} = \frac{x}{50} \quad$ x = 142.4 mg/L as $CaCO_3$

Mg^{++} 48 mg/L $\quad \frac{48}{12} = \frac{x}{50} \quad$ x = 200 mg/L as $CaCO_3$

Na^{+} 26.5 mg/L $\quad \frac{26.5}{23} = \frac{x}{50} \quad$ x = 57.6 mg/L as $CaCO_3$

HCO_3^- as $CaCO_3$ $\quad = \quad$ 240 mg/L as $CaCO_3$

$SO_4^=$ 154 $\quad \frac{154}{48} = \frac{x}{50} \quad$ 160 mg/L as $CaCO_3$

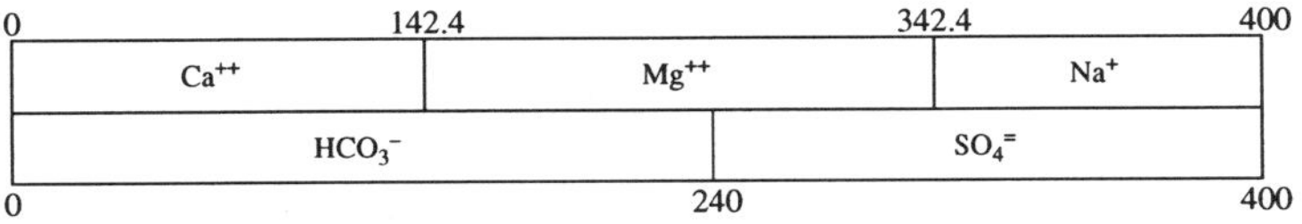

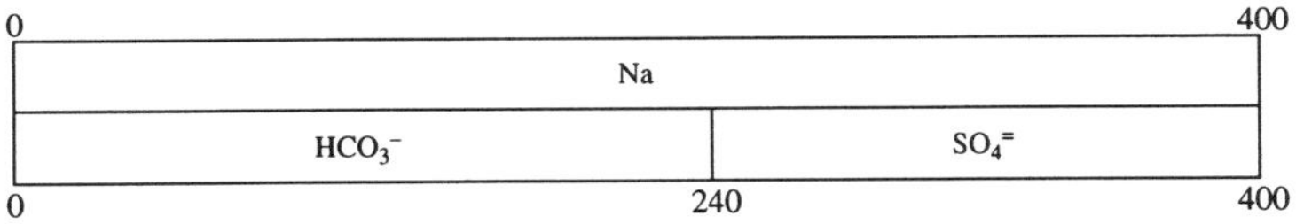

0 – 142.4	142.4 – 342.4	342.4 – 400
Ca^{++}	Mg^{++}	Na^{+}

0 – 240	240 – 400
HCO_3^-	$SO_4^=$

0 – 400
Na

0 – 240	240 – 400
HCO_3^-	$SO_4^=$

$$\frac{342.4 \text{ mg/L Total Hardness (TH)}}{17.12 \text{ mg/L/gr/gal}} = 20 \text{ gr/gal T.H.}$$

$$\frac{28{,}000 \text{ gr/ft}^3 \times 15 \text{ ft}^3}{20 \text{ gr/gal}} = 21{,}000 \text{ gal treated per cycle}$$

Regeneration wastes = 840 gal

With 4% conc. factor is 25

Ca^{++} $142.4 \times 25 + 142.4 = 3702.4$ mg/L Ca as $CaCO_3$

Mg^{++} $200 \times 25 + 200 = 5200$ mg/L Mg as $CaCO_3$

HCO_3^- 240 mg/L as $CaCO_3$

$SO_4^=$ 160 mg/L as $CaCO_3$

150 lb NaCl $\frac{150}{58.5} = \frac{x}{50} \quad$ x = 128 lb NaCl as $CaCO_3$

$$128 = x \times 8.34 \times 0.00084$$

$$x = 18,271 \text{ mg/L NaCl as } CaCO_3$$

and

$$Na^+ \ 18,271 \text{ mg/L as } CaCO_3 - (3702.4 + 5200) = 9768.6$$

$$Cl^- \ 18,271 \text{ mg/L as } CaCO_3$$

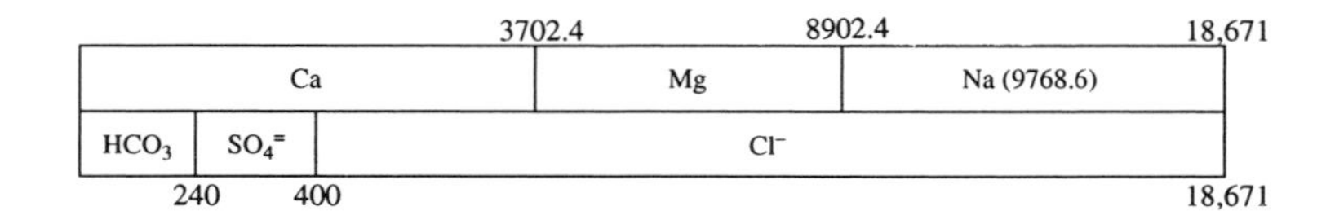

EXAMPLE PROBLEM

A deionization system contains 12 ft³ of cation resin and 10 ft³ of anion resin. How many pounds of H_2SO_4 and NaOH would be required to regenerate the system on a theoretical basis? The capacity of the cation resin is 18 Kgr/ft³ and the anion resin 17 Kgr/ft³, both expressed as $CaCO_3$.

SOLUTION

$$H_2SO_4 = CaCO_3$$

$$\frac{98}{x} = \frac{100}{18 \text{ Kgr/ft}^3 \times 12 \text{ ft}^3}$$

$$x = 211.68 \text{ Kgr } H_2SO_4$$

$$211.68 \text{ Kgr} \div 7 \text{ Kgr/lb} = 30.24 \text{ lb } 100\% \ H_2SO_4$$

$$2NaOH = CaCO_3$$

$$\frac{2 \times 40}{y} = \frac{100}{10 \text{ ft}^3 \times 17 \text{ Kgr/ft}^3}$$

$$y = 136 \text{ Kgr NaOH}$$

$$136 \text{ Kgr} \div 7 \text{ Kgr/lb} = 19.43 \text{ lb } 100\% \text{ NaOH}$$

EXAMPLE PROBLEM

A raw water has been analyzed and found to contain the following:

pH = 7.4
Ca^{++} = 30 mg/L as Ca
Mg^{++} = 36 mg/L as Mg
Na^{+} = 46 mg/L as Na

Alk = 270 mg/L as $CaCO_3$-HCO_3^-
$SO_4^=$ = 106 mg/L as $SO_4^=$
Cl^- = 85 mg/L as Cl^-

A company plans to use this raw water to produce two water sources as follows (not at the same time):

Source #1	30 mg/L total hardness as $CaCO_3$
	40 mg/L alk as $CaCO_3$
	Rest of cations and anions are of no concern
Source #2	Zero total hardness
	70 mg/L alk as $CaCO_3$
	60 mg/L of $SO_4^=$ as $CaCO_3$

The company has a sodium cycle softener, a hydrogen cycle cation exchanger, and a hydroxyl cycle anion exchanger. The system is piped so that water can go through or around the system. Flow can also bypass the hydroxyl cycle anion exchanger. They desire 100,000 gal/day of treated water. The water is stored in open tanks.

1. Draw a flow diagram of the system.
2. Show the flows through each unit for both water sources.
3. Draw a bar diagram for each finished water source.

SOLUTION

$$\text{Ca} = 80 \times \frac{50}{20} = 200 \text{ mg/L as } CaCO_3$$

$$\text{Mg} = 36 \times \frac{50}{12} = 150 \text{ mg/L as } CaCO_3$$

$$\text{Na} = 46 \times \frac{50}{23} = 100 \text{ mg/L as } CaCO_3$$

$$\text{K} = 39 \times \frac{50}{39} = 50 \text{ mg/L as } CaCO_3$$

$$HCO_3 = 270 \text{ mg/L as } CaCO_3$$

$$SO_4 = 106 \times \frac{50}{48} \qquad = \quad 110 \text{ mg/L as } CaCO_3$$

$$Cl \;= 85 \times \frac{50}{35.5} \qquad = \quad 120 \text{ mg/L as } CaCO_3$$

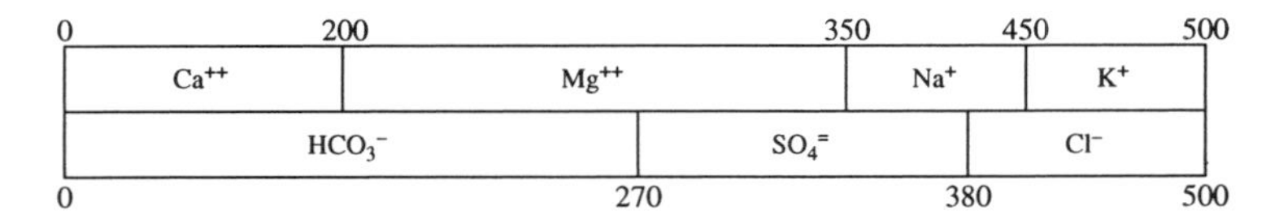

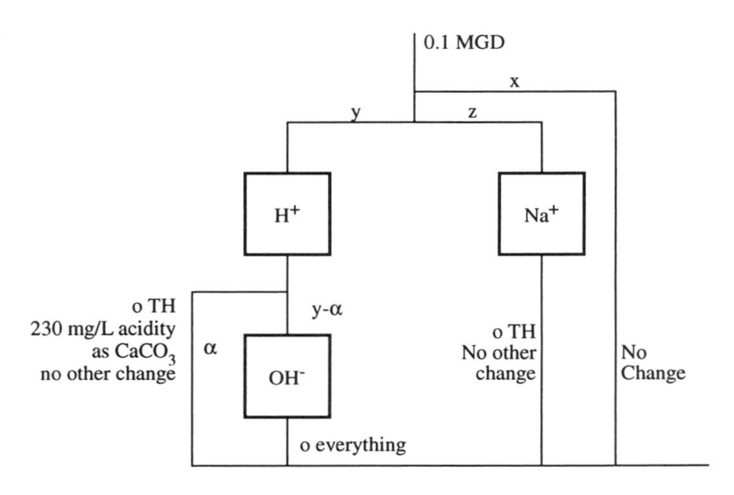

Source 1: 30 mg/L = TH, 40 mg/L = alk

$$TH = 350 \text{ mg/L} \times x = 0.1 \times 30 \text{ mg/L}$$

$$x = 0.0086 \text{ MGD}$$

$$alk = -230 \text{ mg/L} \times \alpha + 0 + 270 \text{ mg/L} \times z + 270 \text{ mg/L} \times 0.0086 = .1 \times 40$$

$$-230\ \alpha + 270\ z = 1.678$$

$$SO_4 = 110 \text{ mg/L} \times \alpha + 0 + 110 \text{ mg/L} \times z + 110 \text{ mg/L} \times 0.0086 = .1 \times 110$$

$$110\ \alpha + 110\ z = 10.094$$

$$2.09 \times SO_4 \quad 230\ \alpha + 230\ z = 21.096$$

$$500\ z = 22.774$$

$$z = 0.0455 \text{ MGD}$$

$$y = 0.0459 \text{ MGD}$$

$$\alpha = 0.0459 \text{ MGD}$$

$$y - \alpha = 0$$

No treatment 0.0086

$Ca = 200 \quad Mg = 150 \quad Na = 100 \quad K = 50$

$HCO_3 = 270 \quad SO_4 = 110 \quad Cl = 120$

Na cycle 0.0455

$Na = 450 \quad K = 50$

$HCO_3 = 270 \quad SO_4 = 110 \quad Cl = 120$

H^+ cycle 0.0459

$H^+ = 500$

$HCO_3 = 270 \quad SO_4 = 110 \quad Cl = 120$

$$Na^+ = .0086 \times 100 + .0455 \times 450 + 0 = .1 \times x$$

$$.86 + 20.475 = .1\ x$$

$$x = 213 \text{ mg/L as } CaCO_3$$

$$Ca^{++} = .0086 \times 200 + 0 + 0 = .1 \times x$$

$$x = 17 \text{ mg/L as } CaCO_3$$

$$Mg^{++} = .0086 \times 150 + 0 + 0 = .1 \times x$$

$$x = 13 \text{ mg/L as } CaCO_3$$

$$K^+ = .0086 \times 50 + 0.0455 \times 50 + 0 = 0.1 \times x$$

$$0.43 + 2.27 = 0.1\ x$$

$$x = 27 \text{ mg/L as } CaCO_3$$

$$HCO_3^- = 0.0086 \times 270 + 0.0455 \times 270 + 0.0459 \times 270 = 0.1 \times x$$

$$x = 270 \text{ mg/L as } CaCO_3$$

$$H^+ = 0 + 0 + 0.0459 \times 500 = 0.1 \times x$$

$$x = 229.5 \text{ mg/L as } CaCO_3$$

$$SO_4^{=} = 0.0086 \times 110 + 0.0455 \times 110 + 0.0459 \times 110 = 0.1 \times x$$

$$x = 110 \text{ mg/L as } CaCO_3$$

$$Cl^- = 0.0086 \times 120 + 0.0455 \times 120 + 0.0459 \times 120 = 0.1 \times x$$

$$x = 120 \text{ mg/L as } CaCO_3$$

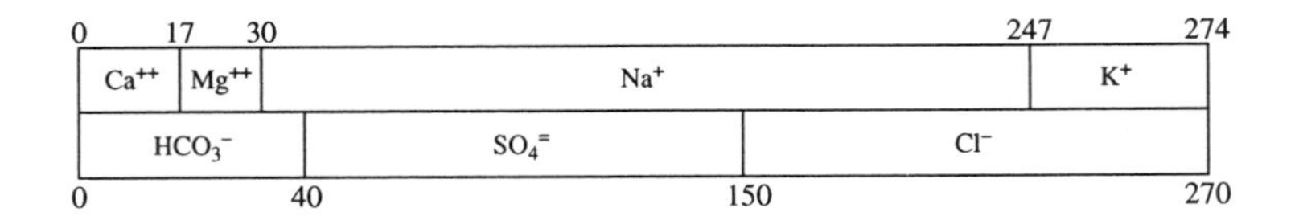

Source 2: TH = 0, alk = 70 mg/L as $CaCO_3$, $SO_4^=$ = 60 mg/L as $CaCO_3$

$$-230\,\alpha + 270\,z = 70 \times 0.1 \text{ alk}$$

$$2.09 \times 110\,\alpha + 110\,z = 60 \times 0.1\ SO_4^=$$

$$230\,\alpha + 230\,z = 125.5 \times 0.1$$

$$500\,z = 195.5 \times 0.1$$

$$z = 0.0391 \text{ MGD}$$

$$110\,\alpha + 110 \times 0.0391 = 60 \times 0.1$$

$$110\,\alpha = 1.699$$

$$\alpha = 0.0154 \text{ MGD}$$

$\alpha = 0.0154$ MGD H^+ cycle

$z = 0.0391$ MGD Na^+ cycle

$y - \alpha = 0.0455$ MGD $H^+ + OH^-$ cycle

$y = 0.0609$ MGD

H^+ cycle 0.0154

$H^+ = 500$ $HCO_3^- = 270$ $Cl^- = 120$ $SO_4^= = 110$

Na^+ cycle 0.0391

$Na^+ = 450$ $K^+ = 50$

$HCO_3^- = 270$ $Cl^- = 120$ $SO_4^= = 110$

$H^+ + OH^- = 0$

$$Na^{+} = 0.0391 \times 450 = 0.1 \times x$$

$$x = 176 \text{ mg/L as } CaCO_3$$

$$K^{+} = 0.0391 \times 50 = 0.1 \times x$$

$$x = 20 \text{ mg/L as } CaCO_3$$

$$H^{+} = 0.0154 \times 500 = 0.1 \times x$$

$$x = 77 \text{ mg/L as } CaCO_3$$

$$HCO_3^{-} = 0.0391 \times 270 + 0.0154 \times 270 = 0.1\ x$$

$$x = 147 \text{ mg/L as } CaCO_3$$

$$SO_4^{=} = 0.0391 \times 110 + 0.0154 \times 110 = 0.1\ x$$

$$x = 60 \text{ mg/L as } CaCO_3$$

$$Cl^{-} = 0.0391 \times 120 + 0.0154 \times 120 = 0.1\ x$$

$$x = 65 \text{ mg/L as } CaCO_3$$

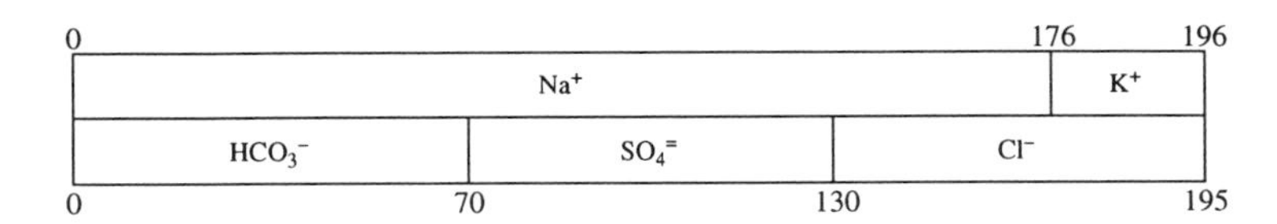

Alternate Solution for Source 1

$$TH = 350 \text{ mg/L} \times x = 0.1 \times 30 \text{ mg/L}$$

$$x = 0.0086 \text{ MGD}$$

$$alk = 0 \times y + 270 \text{ mg/L} \times z + 270 \text{ mg/L} \times 0.0086 = 0.1 \times 40 \text{ mg/L}$$

$$z = 0.0063 \text{ MGD}$$

$$0.1 - (0.0086 + 0.0063) = y$$

$$y = 0.0851 \text{ MGD}$$

No treatment 0.0086 MGD

$Ca = 200$ $Mg = 150$ $Na = 100$ $K = 50$

$HCO_3 = 270$ $SO_4 = 110$ $Cl = 120$

Na cycle 0.0063 MGD

$Na = 450 \qquad K = 50$

$HCO_3 = 270 \qquad SO_4 = 110 \qquad Cl = 120$

$H^+ + OH^-$ cycle 0.0851 MGD

zero for all elements.

$$Na^+ = 0.0086 \times 100 + 0.0063 \times 450 = 0.1 \times x$$

$$x = \frac{2.835 + .86}{0.1}$$

$$x = 37\ mg/L$$

$$Ca^+ = 0.0086 \times 200 = 0.1\ x$$

$$x = 17.2\ mg/L$$

$$Mg^+ = 0.0086 \times 150 = 0.1\ x$$

$$x = 12.9\ mg/L$$

$$K^+ = 0.0086 \times 50 + 0.0063 \times 50 = .1\ x$$

$$x = \frac{0.43 + 0.315}{.1} = 7.5\ mg/L$$

$$HCO_3^- = 0.0086 \times 270 + 0.0063 \times 270 = 0.1 \times\ x$$

$$x = \frac{2.32 + 1.70}{.1} = 40\ mg/L$$

$$SO_4 = 0.0086 \times 110 + 0.0063 \times 110 = 0.1 \times\ x$$

$$x = \frac{0.946 + 0.693}{0.1} = 16\ mg/L$$

$$Cl = 0.0086 \times 120 + 0.0063 \times 120 = 0.1 \times\ x$$

$$x = \frac{1.032 + 0.756}{0.1} = 18\ mg/L$$

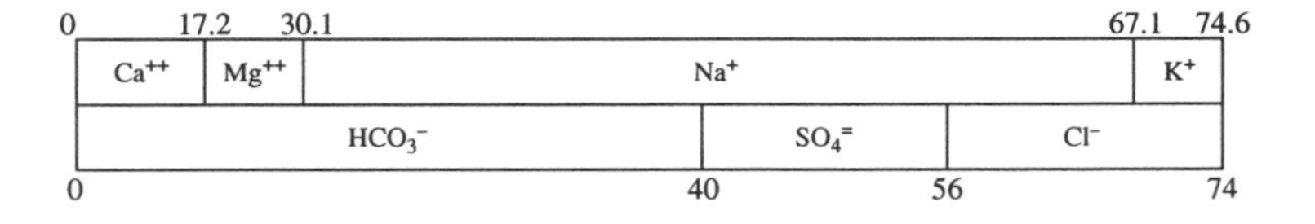

SOLUTION 3

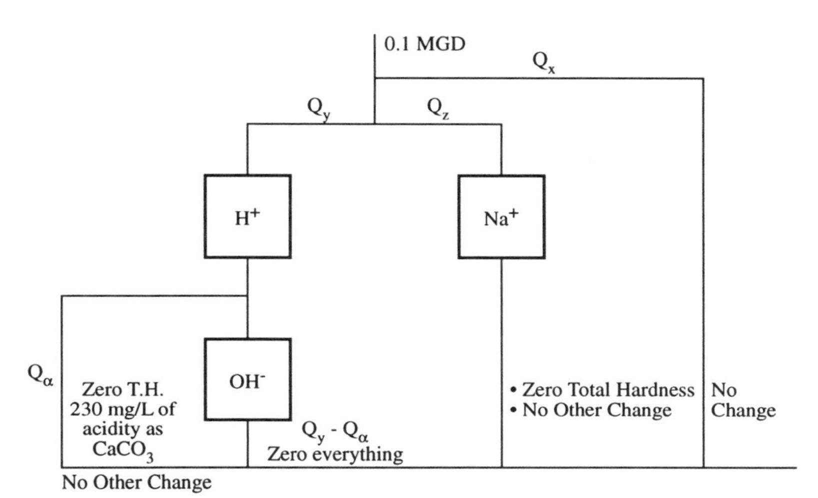

Source 1

$$\text{Total hardness} = (350 \text{ mg/L})\ Q_z = (0.1)(30 \text{ mg/L})$$

$$Q_z = 0.0086 \text{ MGD}$$

$$\text{Alkalinity} = (-230 \text{ mg/L})(Q_x) + 0 + (270 \text{ mg/L})\ Q_z + (270 \text{ mg/L})(0.0086) = (0.1)(40 \text{ mg/L})$$

$$(-230 \text{ mg/L})(Q_x) \times (270)\ Q_z = 1.678$$

$$SO_4 = (110 \text{ mg/L})\ Q_\alpha + 0 + (110 \text{ mg/L})\ Q_z + (110 \text{ mg/L})(0.0086) = 0.1 \times 110$$

$$110\ Q_\alpha + 110\ Q_z = 10.094$$

$Q_x = 0.0086$ MGD

$Q_z = 0.0455$ MGD

$Q_y = 0.0459$ MGD

$Q_\alpha = 0.0459$ MGD

No treatment line, Q = 0.0086 MGD

$Ca = 200$ $Na = 100$ $HCO_3 = 270$ $Cl = 120$

$Mg = 150$ $K = 50$ $SO_4 = 110$

Na^+ cycle		H^+ cycle	
Q_z	= 0.0455 MGD	Q_y	= 0.0459 MGD
Na^+	= 450 mg/L	H^+	= 500 mg/L
K^+	= 50 mg/L	HCO_3^-	= 270 mg/L
HCO_3^-	= 270 mg/L	$SO_4^=$	= 110 mg/L
$SO_4^=$	= 110 mg/L	Cl^-	= 120 mg/L
Cl^-	= 120 mg/L		

$$Na^+ = (0.0086)(100) + (0.0455)(450) + 0 = (0.1)\ x$$

$$x = 213 \text{ mg/L as } CaCO_3 = Na^+$$

$$Ca^{++} = (0.0086)(200) + 0 + 0 = (0.1)(Ca^{++})$$

$$Ca^{++} = 17 \text{ mg/L as } CaCO_3$$

$$Mg^{++} = (0.0086)(150) + 0 + 0 = (0.1)(Mg^{++})$$

$$Mg^{++} = 13 \text{ mg/L as } CaCO_3$$

$$K^+ = (0.0086)(50) + (0.0455)(50) + 0 = (0.1)(K^+)$$

$$K^+ = 27 \text{ mg/L as } CaCO_3$$

$$HCO_3^- = (0.0086)(270) + (0.0455)(270) + (0.0459)(270) = 0.1\ x$$

$$HCO_3^- = x = 270 \text{ mg/L as } CaCO_3$$

$$HCO_3^- = 270 \text{ mg/L as } CaCO_3$$

$$H^+ = 0 + (0.0459)(500) = (0.1)(H^+)$$

$$H^+ = 229.5 \text{ mg/L as } CaCO_3$$

$$Cl^- = (0.0086)(120) + (0.0455)(120) + (0.0459)(120) = (0.1)(Cl^-)$$

$$Cl^- = 120 \text{ mg/L as } CaCO_3$$

Source 1 Bar Chart

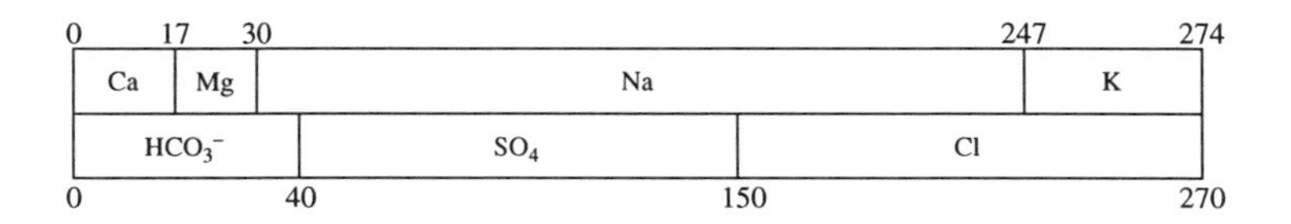

Source 2

- Zero total hardness
- Alkalinity 70 mg/L as $CaCO_3$
- Sulfate 60 mg/L as $CaCO_3$

$$(1)\quad -230\ Q_\alpha + 270\ Q_z = (0.1)(70) \qquad -\text{Alkalinity}$$

$$(2)\quad 110\ Q_\alpha + 110\ Q_z = (0.1)(60)$$

Solve:

$$Q_\alpha = 0.0154\ \text{MGD}, \qquad H^+ \text{ cycle}$$

$$Q_z = 0.0391\ \text{MGD}, \qquad Na^+ \text{ cycle}$$

$$Q_y - Q_\alpha = 0.0455\ \text{MGD}, \qquad H^+ + OH^- \text{ cycle}$$

$$Q_y = 0.0609\ \text{MGD}$$

H^+ cycle, $Q_\alpha = 0.0154$ MGD

$H^+ = 500 \quad HCO_3^- = 270 \quad Cl^- = 120 \quad SO_4^= = 110$

Na^+ cycle, $Q_z = 0.0391$ MGD

$Na^+ = 450 \quad K^+ = 50 \quad HCO_3^- = 270 \quad Cl^- = 120 \quad SO_4^= = 110$

$H^+ + OH^- = 0$

$$Na^+ = (0.0391)(450) = (0.1)(Na^+)$$

$$Na^+ = 176\ \text{mg/L as } CaCO_3$$

$$K^+ = (0.0391)(50) = (0.1)(K^+)$$

$$K^+ = 20\ \text{mg/L as } CaCO_3$$

$$H^+ = (0.0154)(500) = (0.1)(H^+)$$

$$H^+ = 77 \text{ mg/L as } CaCO_3$$

$$HCO_3^- = (0.0391)(270) + (0.0154)(270) = (0.1)(HCO_3^-)$$

$$HCO_3^- = 147 \text{ mg/L}$$

$$SO_4^= = (0.0391)(110) + (0.0154)(110) = 0.1\ SO_4^=$$

$$SO_4^= = 60 \text{ mg/L as } CaCO_3$$

$$Cl^- = (0.0391)(120) + (0.0154)(120) = (0.1)(Cl^-)$$

$$Cl^- = 65 \text{ mg/L as } CaCO_3$$

Source 2 Bar Chart

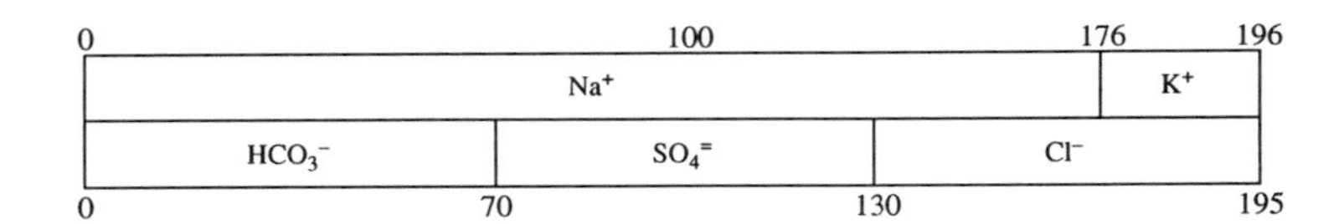

EXAMPLE PROBLEM

You have been asked to check the capacity of a deionization system that is treating a water with 700 mg/L NaCl. The leakage is 2.5 mg/L NaCl. The system contains 13 ft³ of strong acid cation resin with a capacity of 16 Kgr/ft³ and a strong-base anion resin unit containing 12 ft³ with a capacity of 14 Kgr/ft³. Sulfuric acid efficiency is 0.64 lb 56° Bé/Kgr and sodium hydroxide efficiency is 0.51 lb/Kgr. What is the capacity in gallons per cycle? How much additional capacity could they get on the cation unit by using a chemically equivalent amount of HCl? What will be the alkalinity or acidity of the equalized regeneration wastes?

SOLUTION

Cation resin

$$13 \text{ ft}^3 \times 16 \text{ Kgr/ft}^3 = 208 \text{ Kgr as } CaCO_3$$

Anion resin

$$12\ ft^3 \times 14\ Kgr/ft^3 = 168\ Kgr\ as\ CaCO_3$$

$$\frac{700}{58.5} = \frac{x}{50}$$

$$x = 598\ mg/L\ as\ CaCO_3\ or\ 35\ gr/gal\ as\ CaCO_3$$

$$\frac{168,000\ gr/cycle}{35\ gr/gal} = 4800\ gal/cycle$$

$$\frac{208,000\ gr/cycle}{35\ gr/gal} = 5943\ gal/cycle$$

No increase in capacity because no Ca in raw water

Regeneration wastes = 50 to 70 gal H^+ water/ft^3 to rinse anion unit

H_2SO_4 $60 \times 12 = 720$ gal

0.64×0.98 lb/Kgr $720\ gal \times 35\ gr/gal = 25,200\ gr$ or 25.2 Kgr

$$(168 + 25.2)\ 0.64 \times 0.98 = 121.2\ lb\ 100\%\ H_2SO_4$$

NaOH

$$0.51\ lb/Kgr \times 168\ Kgr = 85.7\ lb\ 100\%\ NaOH$$

$$121.2 \times \frac{50}{49} = 123.7\ lb\ H_2SO_4\ as\ CaCO_3\ acidity$$

$$85.7 \times \frac{50}{40} = 107.1\ lb\ NaOH\ as\ CaCO_3\ alk$$

$$7\ Kgr/lb \times 123.7 = 865.9\ Kgr\ H_2SO_4\ as\ CaCO_3$$

$$7\ Kgr/lb \times 107.1 = 749.7\ Kgr\ NaOH\ as\ CaCO_3$$

865.9 Kgr H_2SO_4 as $CaCO_3$ put into cation unit
−193.2 Kgr H_2SO_4 as $CaCO_3$ used in resin
672.7 Kgr H_2SO_4 as $CaCO_3$ not used (in regen.)

749.7 Kgr NaOH as $CaCO_3$ put into anion unit
−168.0 Kgr NaOH as $CaCO_3$ used in resin
581.7 Kgr NaOH as $CaCO_3$ not used (in regen.)
− 25.2 Kgr H_2SO_4 as $CaCO_3$ from H^+water rinse
556.5 Kgr NaOH as $CaCO_3$ not used

$$\begin{array}{r} 672.7 \\ -\underline{581.7} \\ 91.0 \end{array} \text{ Kgr } H_2SO_4 \text{ as } CaCO_3 \text{ acidity}$$

or

$$\begin{array}{r} 672.7 \\ -\underline{556.5} \\ 116.2 \end{array} \text{ Kgr } H_2SO_4 \text{ as } CaCO_3 \text{ acidity}$$

With 4% NaOH

8% H_2SO_4

$$\frac{85.7}{.04 \times 8.34} = 257 \text{ gal}$$

$$\frac{121.2}{.08 \times 8.34} = \frac{182 \text{ gal}}{439 \text{ gal}}$$

$$\frac{91,000 \text{ gr}}{439} = 207.3 \text{ gr/gal acidity as } CaCO_3 \quad \text{or} \quad 3549 \text{ mg/L}$$

$$\frac{116,200 \text{ gr}}{439 + 720} = 100.3 \text{ gr/gal acidity as } CaCO_3 \quad \text{or} \quad 1716 \text{ mg/L}$$

REFERENCES

Anderson, R. E., "Ion-Exchange Separations," in *Handbook of Separation Techniques for Chemical Engineers*, McGraw-Hill, New York, 1979.

Applebaum, S. B., *Demineralization by Ion Exchange in Water Treatment and Chemical Processing of Other Liquids*, Academic Press, New York, 1968.

Etzel, J. E. and Keramida, V., "Treatment of Metal Plating Wastes with an Unexpanded Vermiculite Exchange Column," U.S. Patent No. 13,929, 1979.

Helfferich, F., *Ion Exchange*, McGraw-Hill, New York, 1962.

Kunin, R., *Ion Exchange Resins*, 2nd ed., Wiley, New York, 1958.

Nachod, F. C. and Shubert, J., Eds., *Ion Exchange Technology*, Academic Press, New York, 1956.

Rohm and Haas Company, Amberlite® IR-120 Plus Specification Sheet, January 1982, Philadelphia.

Weber, W. J., Jr., *Physiochemical Processes for Water Quality Control*, Wiley-Interscience, New York, 1972.

CHAPTER 4

Laboratory-Scale Testing of Ion Exchange

This chapter provides the tools to make ion exchange work for you. It describes the procedures and components needed to perform testing of columnar ion exchange systems.

Bench-scale ion exchange systems can be used to do a myriad of things ranging from a simple experiment to determine if a particular ion exchange resin in a particular ionic form will remove a particular target ion from a wastewater stream, to developing complete design data for a new ion exchange system. It is often used to develop data for graduate degrees. It is also used to test or evaluate the performance of a newly developed resin. The techniques are basically the same and modified as required.

Other sources (Diamond Shamrock Chemical Co., 1969; Anderson, 1979) describe the necessary hardware and components for designing small-scale testing equipment. They may offer additional helpful hints and advice to make preliminary studies, a relatively easy way to properly design a full-scale unit. We highly recommend the *Amberlite® Ion Exchange Resins Laboratory Guide*, published in 1979 by the Rohm and Haas Company; this is an excellent reference on the subject.

INTRODUCTION

The absolute beauty of designing ion exchange systems is that scale-up is governed mainly by the diffusion of ions into and out of the ion exchange beads and not by the usual hydraulic considerations governing filtration scale-up. What this means is that design questions can be answered quickly and with little cost.

Often times we need to know qualitatively if the resin we have selected, in the ionic form in which we have chosen to operate the resin, will remove a particular target ion. In this chapter, you will learn that these data are often gathered using small-diameter titration burettes containing less than a foot of resin depth.

Today, respirometry is often used to predict the performance of specified treatment systems in treating the waste. The heavy metals that inhibit biological growth can be removed using a bench-scale ion exchange system, allowing a quick determination of the effect of that ion on the process.

COLUMNS

Laboratory (bench)-scale ion exchange columns are usually glass, heavy-walled polyethylene, tygon (i.e., polyvinyl chloride) tubes or pipes of various lengths. For preliminary go–no-go type studies, we have even used syringes. Be innovative.

The resin or exchanger is supported on a plug of glass wool, glass wool/glass beads, glass beads, or a sintered glass disc fitted into the column near the bottom. Our experience suggests that glass wool plugs and small glass beads or a small-diameter screen placed over the rubber stopper offer the best support for columns up to 1 in. in diameter. Glass discs can clog and thus your evaluation is interrupted until you either replace the disc or put together another column. Glass wool plugs can be tricky when used in burettes. Glass beads work well in almost all applications. Figures 4-1 through 4-8 show various column designs and components.

RESERVOIRS

Any convenient vessel can be used to hold and feed the wastewater, regenerant, backwash, and rinse solutions. For bench-scale gravity set-ups our experience is that a separatory funnel or funnels is the most effective reservoir. Figure 4-1 shows an option where the separatory funnel is connected to the ion exchange column via a stopper. Flow rates are controlled by adjusting the stopcock in the separatory funnel and the screw clamp located at the bottom of the column. A graduated cylinder and stopwatch are used to determine solution flow rates. Effluent samples are collected from the glass tip. Flow rates are maintained by adding solution to the top of the funnel. Because flow rates are low, usually in the milliliters per minute range, automation and pumps are not required.

A modification of the scheme can be used to feed a burette. Figure 4-5 shows the set-up.

Figure 4-6 shows the burette set-up of Figure 4-5 automated with a positive displacement pump. The feed vessel in this case is a 4-L capacity glass jug. This particular system is operated in an upflow mode. The same set-up can be used in a downflow mode.

COLUMN OPERATIONS

Before we actually use or operate our ion exchange column, we must consider and determine the flow rates we intend to use for each cycle and the concentration and volume of regenerant we intend to use.

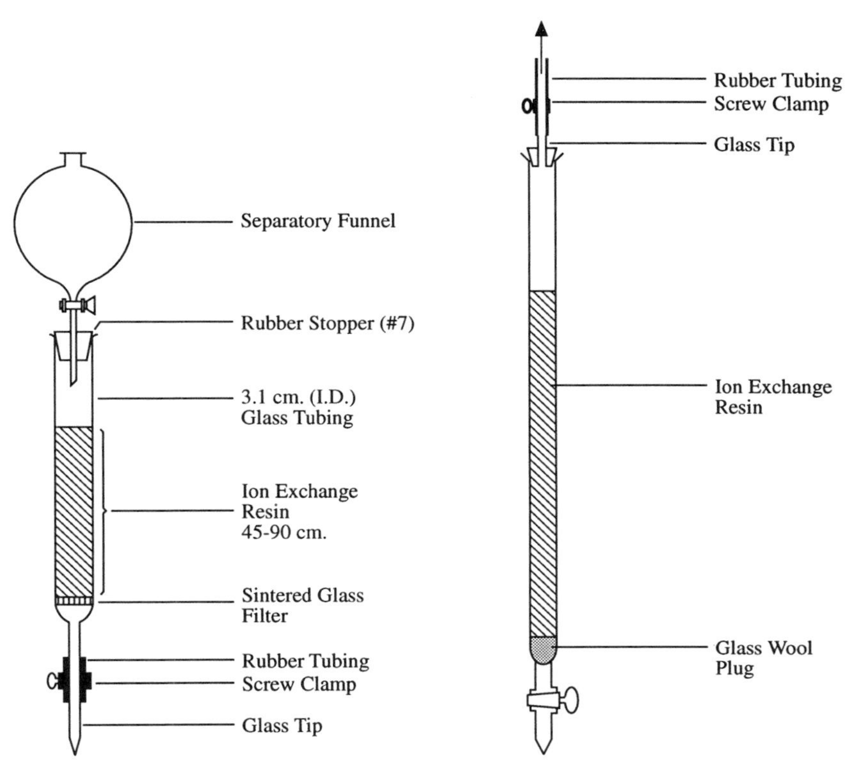

Figure 4-1 Simple gravity-flow burette set-up with sintered glass filter.

Figure 4-2 Typical gravity-flow burette set-up with glass-wool plug.

Flow Rates

Flow rates are important in each ion exchange cycle. They are expressed in two ways: volumetric flow rate (volume velocity), in which the units are gpm/ft^3, bed volumes (BV)/hr or L/L, and surface loading rate (linear velocity), in which the units are gpm/ft^2 or m^3/m^2-hr (m/hr). Volumetric flow rates are linear velocity (ft/min) divided by bed-height (ft).

$$\frac{\text{ft}}{\text{min}} \times \frac{7.48 \text{ gal}}{\text{ft}^3} = \frac{\text{gal}}{\text{min ft}^2}$$

$$\frac{\text{gal}}{\text{min ft}^2} \times \frac{1}{\text{ft}} = \frac{\text{gal}}{\text{min ft}^3}$$

The term bed volume (BV) is the volume of resin including voids in the column.

The resin has been backwashed, settled, placed, regenerated to the proper ionic form, and rinsed. Use a precalibrated column or burette to accurately

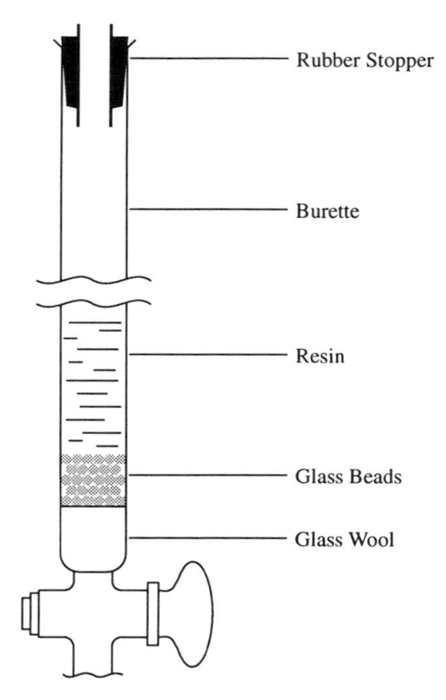

Figure 4-3 Typical gravity-flow burette set-up with glass-wool plug and glass beads.

measure the volume. If this is not possible, drain the resin into a graduated cylinder and measure the settled resin after the laboratory evaluation is complete.

Volumetric flow rate is inversely related to contact time between a solution and the ion exchange resin (exchanger) and is the flow rate(s) used during the preliminary stages of evaluation. The surface loading rate relates directly to scale-up and must be considered during larger column studies involving resin attrition or service flow rate performance.

Typical flow rates are as follows (these are guidelines only):

Service cycle: 0.75–1.50 gpm/ft^3 (0.1–0.2 BV/min)
Backwash: Upflow to cause 50–75% bed expansion — see Table 3-6
Regeneration: 0.5–0.75 gpm/ft^3 (0.05–0.1 BV/min)
Rinse: Slow at regeneration rate, fast at service cycle rate

Resin manufacturers always supply recommended flow rate data for their resins. See Table 3-6. For standard applications consult your resin manufacturer. For unique, and many industrial, applications you will probably have to determine the optimum flow rate that satisfies your particular application.

Proper flow rates can be maintained by using gravity and overhead reservoirs and by regulating flow via screw clamps (see Figures 4-1 and 4-2), by using a positive displacement laboratory pump (Figure 4-6), or by using a constant head

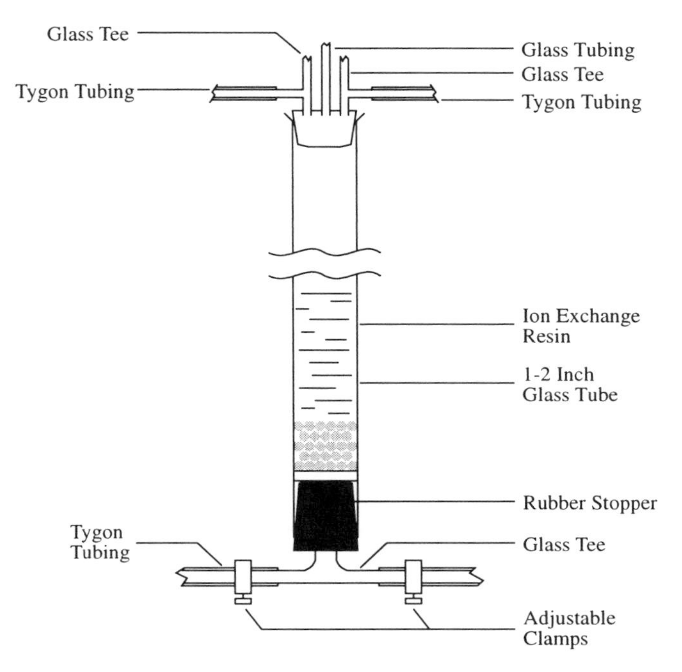

Figure 4-4 Fully configured ion-exchange column (clamps allow all cycles) — gravity or pump.

device (Figure 4-8). The duration of the test, your budget, and the data required will be your guide.

Resin Conditioning

Ion exchange resins are shipped in the more common ionic forms:

Strong-acid resin:	sodium or hydrogen
Strong-base resin:	chloride
Weak-acid resin:	hydrogen
Weak-base resin:	free base

Theoretically, the only conditioning step required is to rinse the resin thoroughly with deionized water and then let it soak to attain its final size. If you require the resin to be in another ionic form, you must regenerate and properly rinse the resin prior to use in the service cycle.

We recommend that you condition a new sample of resin through one or two acid-base cycles to remove any soluble impurities remaining from the manufacturing process. Use a 4% strong acid such as HCI and a 4% strong base such as NaOH to perform the conditioning. Remember that if the resin is in the sodium cycle or free base form, pass the strong acid through first. If the resin is in the hydrogen or chloride form, pass the strong base through first.

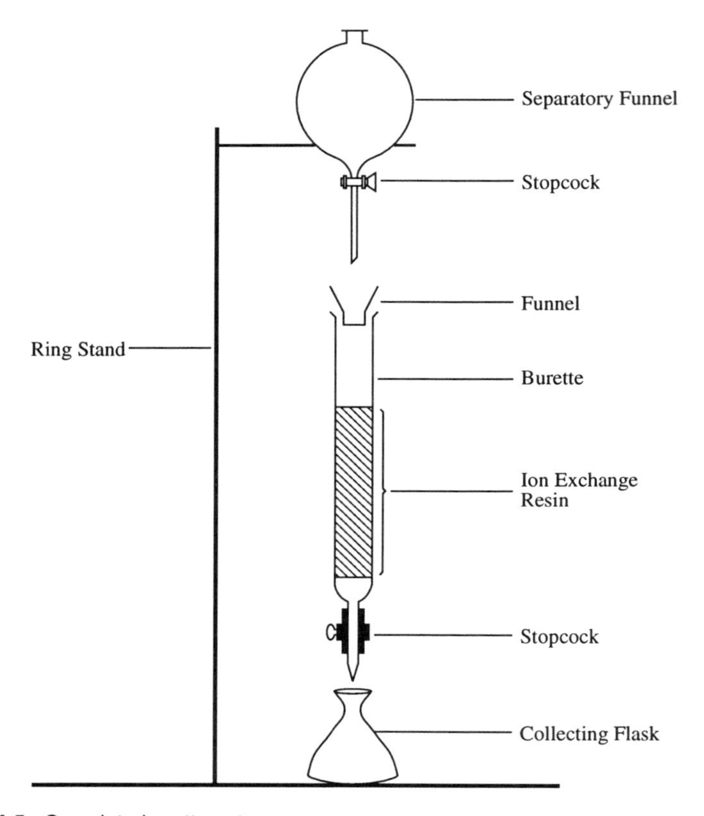

Figure 4-5 Complete burette set-up.

Concentration of Solutions

Absorption step — <2000 mg/L, usually a given
Backwash step — use deionized water for analytical experiments, use actual backwash water for design
Regeneration step — about 1.0 Normal in general; 4–10% usually acceptable (see manufacturer's literature as a starting point)
Rinse step — deionized water

Volumes of Solutions

Service cycle — for quantitative exchange use an amount of solution that contains ions equivalent to half of the ion exchange capacity of the resin as measured in the column, until breakthrough
Backwash — use as much as is required to achieve desired results (see manufacturer's literature)
Regeneration — use amounts calculated to provide from 150 to 500% in excess of theoretical capacity of resin in column, using manufacturer's literature as a base number

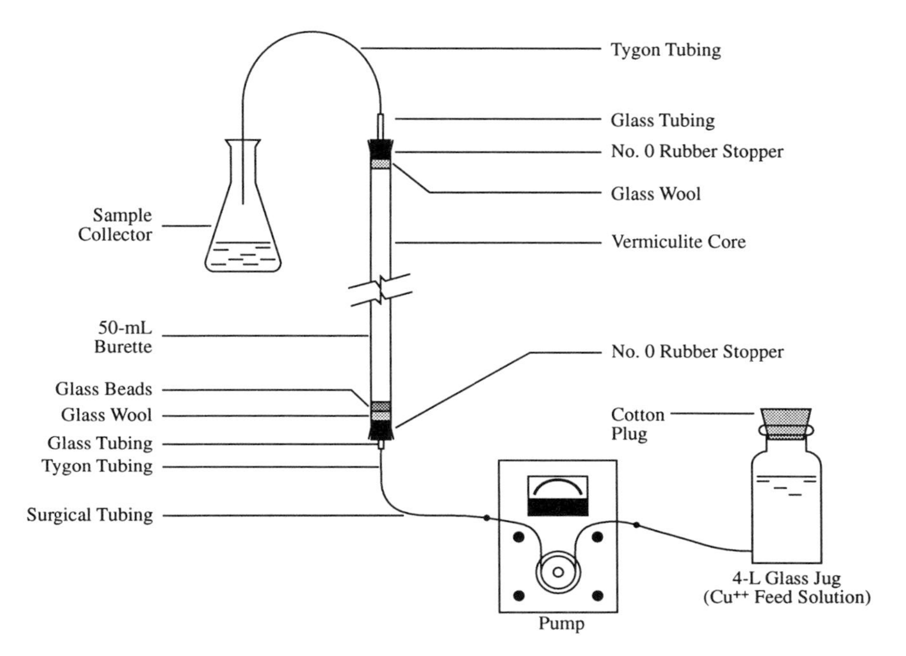

Figure 4-6 Burette set-up with pump.

Rinse — use enough to rinse excess regenerant from the column; usually 10 BVs of deionized water are adequate; check chemically to ensure completeness

Note: About 30–40% of the total volume occupied by an ion exchange resin in a column is void space. The solution you are treating will displace the water filling this space so discard the first BV before collecting the treated solution, *unless* a quantitative experiment is being performed.

Cautions

Keep the following precautions in mind whenever you are operating an ion exchange column.

- Always fully hydrate the ion exchange resin before charging the column.
- Always ensure that the column contains water.
- Never allow the resin to dry. (If the resin does dry out, remove it from the column — rehydration in the column could crack the glass.)
- During the evaluation, always maintain a constant liquid level over the top of the bed.
- Always keep the resin submerged.
- Nitric acid and other strong oxidizing agents can cause explosive reactions when mixed with ion exchange resins.
- Design all equipment to prevent rapid pressure buildup if using nitric acid or strong oxidizing agents. Do your homework.

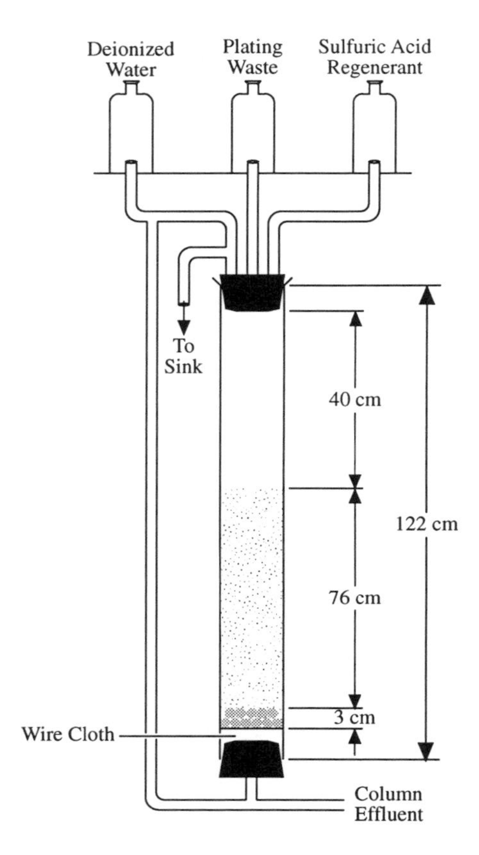

Figure 4-7 Laboratory apparatus for ion exchange experiments.

Charging

Place the resin into a beaker containing deionized water and let the dry resin hydrate; 1 hr is sufficient. Add water to the column equal to one half of the height to be occupied by the resin (e.g., if your resin depth is to be 30 in., add about 15 in. of water). Add the resin-water slurry to the column. Drain any excess water through the bottom of the column, being careful to *never* let the water level fall below the resin. As a rule, load the column with resin to only about one-half of the column height (you will need the remainder of the column for backwashing).

Backwashing

Backwashing a laboratory column is just as important as backwashing a full-scale system. It frees the bed of debris and resin fines, classifies the resin particles, and rids the bed of entrapped air pockets.

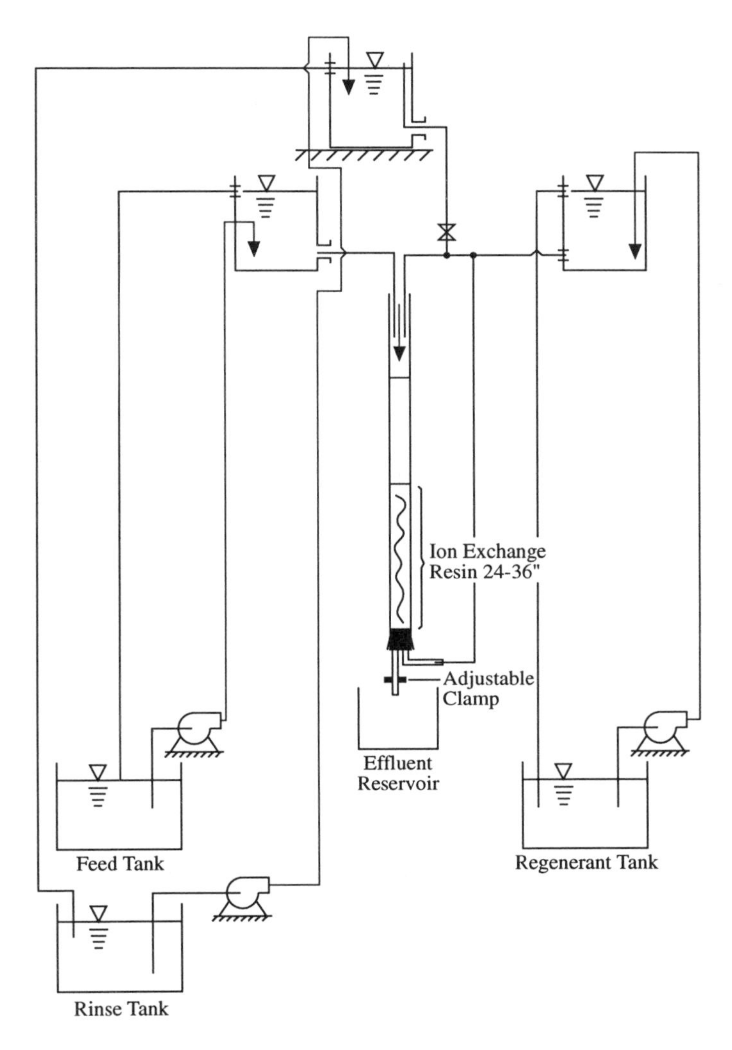

Figure 4-8 Constant head configuration.

1. Attach a water line to the bottom of the column or burette.
2. Very slowly introduce deionized water into the column at a very low flow rate.
3. Increase the flow until the bed of resin expands to near the top end of the column, i.e., is fluidized.
4. Maintain this flow until all air pockets are removed and all the particles have achieved mobility. (Extremely small particles may be allowed to pass out of the column.) This step is most important and, if performed correctly, will result in good particle size classification, with the smaller particles at the top and the larger ones at the bottom portion of the column.
5. Stop the flow of water and permit the resin to settle by gravity.
6. Adjust the liquid level to just above the top of the bed.

Regeneration

To convert an ion exchange resin to the proper ionic form for any reason requires a regeneration step. Regeneration in the laboratory is performed in the same way as regeneration of a full-scale resin. Suitable flow rates were given in the previous section — see the resin spec sheet (Table 3-6). Note that the first BV of rinse water should be applied at the regeneration flow rate because this actually represents the completion of regeneration. (Suggested concentrations of regenerants were also given in the previous section.) In general, concentrations of about 1 to 2 Normal are used (about 4 to 10%). Applied at the suggested flow rates, such concentrations will allow for adequate contact time if the correct volumes of regenerants are used.

Example: The following exchange set-up allows development of design data to treat a plating solution wastewater stream.

Set-up

A schematic diagram of a possible equipment configuration used for ion exchange studies is shown in Figures 4-4 and 4-7. The column consists of a Pyrex tube, 122 cm in length, with an outside diameter of 30 mm and a cross-sectional area of 5.51 cm^2. The column is fitted at both ends with rubber stoppers; on the bottom a No. 5 1/2 and on the top a No. 5. A glass "T" connecting tube is inserted into the bottom stopper and two lengths of glass tubing plus a glass "T" are inserted into the top stopper.

Three reservoirs, each with a 4-L capacity, are placed above the ion exchange column. One reservoir holds the water, wastewater, or solution to be treated and is connected by tygon tubing to one of the lengths of glass tubing inserted in the top stopper. Another reservoir contains the regenerant and is connected to the other glass tube in the top stopper. The third reservoir holds deionized water and is connected to one arm of the glass "T" in the top stopper and also to one arm of the glass "T" in the bottom stopper. To the other arm of the glass "T" in the top stopper a length of tygon tubing is connected. The other end of the tubing is placed in a sink. The other arm of the glass "T" in the bottom stopper is fitted with another length of tygon tubing. All samples are collected through this tubing. All pieces of tygon tubing are fitted with Fisher-Castaloy screw clamps to allow regulation of flow rates.

A piece of 80-mesh stainless steel wire cloth and 3 cm/L of Pyrex glass beads are used as the support for the bed of resin. The resin used in this study is a weak-base anion resin, Amberlite IRA-94. The column contains 420 wet mL of resin, in this case in the sulfate form.

Operating Procedure

The mode of operation is as follows:

1. The wastewater solution is passed through the column in a downflow mode at 1.0 gpm/ft^3 (56 mL/min). Effluent from the column is collected in 100-mL

fractions and placed in small polyethylene containers. The service cycle is continued until the effluent concentration of the target species approaches the influent concentration; i.e., to complete exhaustion.

2. The column is backwashed in an upflow mode with deionized water at a rate sufficient to provide a 50% expansion of the bed of resin. The backwash is continued for approximately 5 min (consult resin specifications for time and flow).
3. The regenerant is passed through the bed in a downflow mode at 0.5 gpm/ft^3 (28 mL/min) (again, refer to resin specifications). Effluent from the column is collected in 100-mL fractions. Regeneration is continued until the major points on a regeneration curve have been defined; i.e., the initial increase in target concentration, the peak concentration, and a characteristic portion of the decrease in target concentration past the peak. Different regenerant concentrations are often used depending on the test objectives. The regenerants are prepared by diluting reagent-grade regenerants with deionized water.
4. The column is then rinsed with deionized water in a downflow mode and the effluent collected in 100-mL fractions. The rinse is continued until the effluent target concentration reaches a predetermined value.

Example: A typical breakthrough curve

Scale-up

Data derived from small-column experiments can be scaled up directly to any diameter column if the height of the bed remains constant. If the small-column experiments were done at a reasonable height (2–3 ft), then increasing this height in a full-scale design usually will not change the shape of the breakthrough curve. In exchanges where the separation factor is greater than 1 for the ion to be removed, the exchange zone (or front) will be relatively small with respect to the column height. Deepening the column in this case should not increase the breakthrough capacity with respect to bed volumes.

Maintaining the same volume flow rate as determined in the small-scale experiments will provide similar cycle times and effluent concentration profiles. If the height of the column is kept the same, then the surface area flow rate will also remain equal. If the column is deepened and the volume flow rate is kept the same, the surface area flow rate will be increased by the proportion that the height has increased. This should not be a problem unless a critical range of flow velocities is reached. Typical linear flow rates are in the range of 4–15 gpm/ft^2. Excessive linear flow rates will greatly increase the pressure drop through the column and could adversely affect the stability of the resin beads.

Once the optimum volume flow rate is known, the actual size of the full-scale contactor can be determined. The amount of resin volume needed to treat a given flow of water will be simply

$$\begin{aligned}\text{Required resin volume} &= \frac{\text{treated water flow rate}}{\text{volumetric flow rate}}\\ &= \frac{\text{gpm}}{\text{gpm/ft}^3}\\ &= \text{ft}^3\end{aligned}$$

Based on this volume and the desired depth of the resin, the diameter of a single column can be determined. Should the required diameter be much larger than 12–15 ft, two or more columns should be used. Typical bed depths used in industry are 2.5–9 ft.

Unless the treated water flow demand is intermittent, interruption of the service cycle for regeneration will require two or more columns or a treated water storage reservoir. If the exhaustion cycle is fairly long (>16–24 hr), a reservoir can provide sufficient water during regeneration time, normally 1–2 hr. The regeneration requirements can be calculated using the manufacturer's design data or laboratory studies. For most ion exchange applications, a typical regeneration cycle is:

1. Backwash
2. Regeneration
3. Slow displacement rinse
4. Fast rinse
5. Service run

Backwashing is typically done to reclassify the resin so that there will be a gradual increase in particle size from top to bottom. This will help to prevent channeling. Ion exchange media will act as good filter media, hence backwashing will remove trapped particulate matter from the resin. Bed expansion in the 50–75% range is normal, and proper freeboard should be allowed for in-column design. This step will last 5–15 min.

Regenerant consumption based on design criteria must be determined per cycle. The rinses following regeneration are normally concurrent: the slow rinse for one to two bed volumes at the regeneration flow rate to displace most of the regenerant from the bed, and the fast rinse at the rate of service flow for 10–30 min. An inventory of wastewater volumes must be calculated to adequately prepare for disposal. This is typically a costly part of operation and maintenance together with regenerant chemical costs. It may be the critical factor in many potential applications where disposal of concentrated brines would be a problem.

REFERENCES

Anderson, R. A., "Ion Exchange Separations," in *Handbook of Separation Techniques for Chemical Engineers*, P. A. Schweitzer Ed., McGraw-Hill, New York, 1979.

Diamond Shamrock Chemical Co., *Duolite Ion Exchange Manual*, Redwood City, CA, 1969.

Rohm and Haas Company, *Amberlite® Ion Exchange Resins Laboratory Guide*, Philadelphia, 1979.

CHAPTER 5

Ion Exchange Applications and Design

This chapter covers a myriad of applications, showing process flow diagrams and numerous example design problems that take the reader from problem conception to the final design. The proof is not left to the student.

We found that the best way to teach a topic is through numerous examples that leave nothing to the imagination. In this chapter we introduce the unique ideas developed by us through the years.

We design ion exchange systems using a few basic "rules of thumb," experience, the resin manufacturer's design recommendations, common sense, and the question, "What do we want to do with this water?" We use bar charts of the water to be treated and draw bar charts of the finished water.

In the applications sections, we provide the basic configurations (ion exchange processes) used today. These flow sheets are general in nature and, when combined with the information provided in the detailed design problems, offer the reader the ability to solve most of the ion exchange problems one might encounter. Additional problems are provided.

THE BASIC FLOW SHEETS

If you look in the ion exchange section of most textbooks and introductory articles on ion exchange, or in the class notes generated by most professors (except the present authors), you will find about ten classic flow sheets. These flow sheets, originated by Applebaum in 1968, are probably the most referenced in the world today. The basic flow sheets present many of the basic principles of ion exchange. They are also an excellent guide for designing ion exchange systems to produce high-quality industrial and potable waters. In all cases, the species being removed are ionized and can effectively be removed by ion exchange. The feed waters are free of suspended solids, oil and grease, and high concentrations of organics, so no pretreatment is required. The level of dissolved solids is well within the economics of efficient ion exchange. No consideration is given to minimizing regenerant or to regenerant disposal. Figure 5-1 is a key showing the basic building blocks.

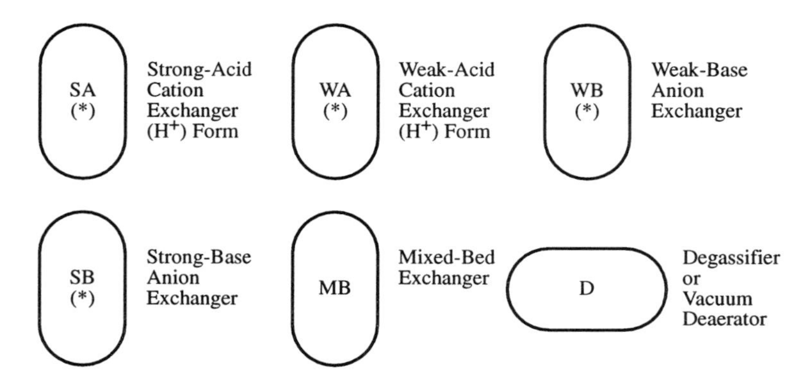

Figure 5-1 Key of basic building blocks. *, Exchange form, H^+, for example. If not indicated, assume (H^+) for SA and WA and (OH^-) for SB and WB.

DEMINERALIZATION

Flow Scheme Configurations

A number of demineralizer flow schemes are available using the following building blocks: strong-acid cation and weak-acid cation exchangers, strong-base anion and weak-base anion exchangers, and a decarbonator or vacuum deaerator (degassifier).

Factors that influence the selection of a demineralizer's flow scheme are raw water composition, effluent water composition (more specifically the necessity of silica removal and alkalinity reduction), plant size, and regenerant cost.

Silica Removal

If silica removal is not required for a particular application then a weak-base resin can be employed. Recall that a weak-base resin has about double the capacity for removing total mineral acidity (sulfates and chlorides) of a strong-base resin, given equal caustic soda regenerant levels. Moreover, it can be regenerated with soda ash, both of which are less expensive than caustic soda. Conversely, if the demineralized water is used for boiler feed or some other industrial purpose in which the silica tolerance is low, a strong-base anion resin must be used.

Degasification

Several materials (that are not ionized) *must* be removed from boiler water. Oxygen must be removed or there are massive corrosion problems. Dissolved CO_2 must also be minimized to stop corrosion problems.

By using the second flow scheme the alkalinity will be boiled out in the vacuum degassifier. Thus the anion unit can be a weak-base anion unit. Chemical regenerant costs can be minimized because the alkalinity does not have to be exchanged. It is also easier to get the CO_2 out in this configuration because the cation effluent is in the acid range. Strong-base anion resin costs more than

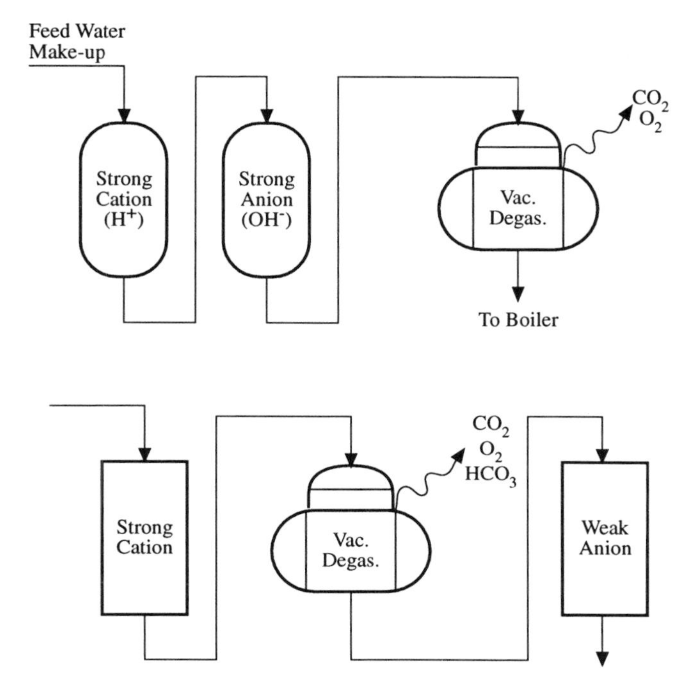

Figure 5-2 Degassification in demineralization.

weak-base resin and the strong-base resin requires a higher concentration of caustic for regeneration. This system does not require salt-splitting ability.

As a rule of thumb, mechanical reduction of carbon dioxide is usually more economical than its chemical removal by the strong-base anion exchanger when the influent alkalinity exceeds 50 to 100 mg/L (as $CaCO_3$) and the demineralizer capacity exceeds 100 gpm.

Heavy Raw Water Composition

A raw water analysis is used to calculate the proportion of sodium and alkalinity (expressed in percent of total cations and anions) and the proportion of total mineral acidity (sulfates and chlorides) and silica, expressed in percent of total exchangeable anions. Higher percentages of sodium and lower percentages of alkalinity increase cation leakage and require higher acid levels to keep this leakage within the limits set for total dissolved solids (TDS). If the acid levels become excessive, four-bed or combined two-bed and mixed-bed demineralizers are recommended. Waters that have high percentages of alkalinity make use of weak-acid cation resins' economy because it saves acid; for such waters a decarbonator or vacuum deaerator becomes advisable (see Figure 5-2). Waters that have high percentages of silica cause much silica leakage and require high caustic soda levels; therefore, strong-base anion resins in both the primary and secondary stages are recommended, to save caustic soda. If the percentage of total mineral acidity is high enough, a weak-base anion resin in a three-bed or double-layer system is justified, to reduce the caustic soda requirements.

Plant Size

The larger the plant, the greater the importance of reducing the operating cost, even though the capital investment is greater. In large plants it is possible to justify the selection of four-bed instead of two-bed demineralizers and the inclusion of both weak-acid cation resin and weak-base anion resin by the savings in regenerant cost. In small plants the small capital investment is usually the controlling factor, and for this reason mixed beds alone are often preferred, even though the cost of the regenerant is high.

Cost of Regenerants

Hydrochloric acid as a regenerant enhances the capacity of the cation exchanger in comparison with other acids and prevents fouling of the resin by calcium sulfate during regeneration (see Chapter 3). Therefore, despite its greater cost, it is often preferred for the small demineralizer.

Effluent Water Composition

Table 5-1 gives the range of effluent TDS and silica normally obtainable, with the major silica-removing demineralizer systems treating raw waters of moderate salinity (less than 500 mg/L TDS).

From this table it is evident that "pure" demineralized water (up to 100,000 ohms) is readily obtained from two-bed systems, but that if "very pure" water (up to 10^6 ohms) is required, the use of four-bed or mixed-bed systems is indicated, and if "ultrapure" water (up to 10^7 ohms) is required, the combination two-bed or three-bed plus mixed-bed system must be used. Similarly, when effluent silica is considered, two beds may be expected to reduce the silica to about 0.22 to 0.1 mg/L. If, however, greater silica removals are required, one of the other contacting systems must be employed.

FLOW SCHEMES

Flow Scheme: Two-Bed, Strong Acid, Weak Base

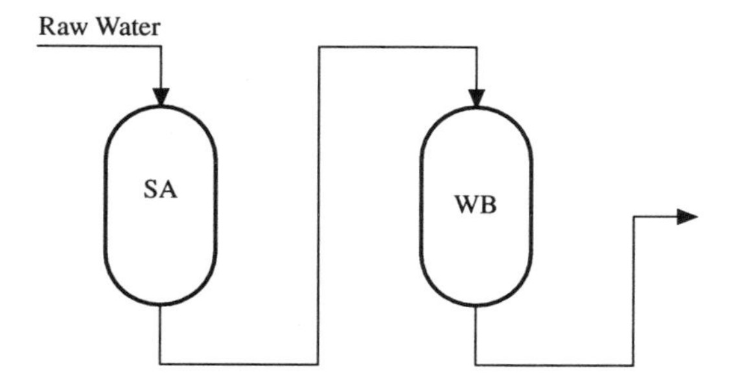

Silica and carbon dioxide not objectionable; quality: 10–30 μmho/cm^3; silica unchanged.

Table 5-1 Range of Electrolyte Content (as Total Dissolved Solids) and Silica in Effluents of Some Major Silica-Removing Demineralizer Systems

	Two- or three-bed	Four-bed	Mixed-bed	Two- or three-bed plus mixed-bed
Electrolyte (as TDS), ppm	2.0–3.0	0.2–1.0	0.2–0.5	0.04–0.10
Silica (as SiO_2),[a] ppm	0.02–0.1	(a) 0.02–0.1 (b) 0.01–0.05	0.02–0.1	(a) 0.02–0.1 (b) 0.01–0.05
Conductivity, μmho	10.0–15.0	1.0–5.0	0.5–1.25	0.10–0.25
Specific resistance, ohms-cm	67,000–100,000	200,000–1,000,000	800,000–2,000,000	4,000,000–10,000,000

[a] The two ranges given signify (a) with primary weak-base anion resin and (b) with primary strong-base anion resin.

Note: This table is based on the demineralization of waters of moderate salinity less than about 500 ppm TDS. Condensate high-rate mixed-bed demineralizers produce effluent TDS of less than 0.04 ppm, or 0.1 μmho (10,000,000 ohms), and SiO_2 less than 0.005 ppm.

Applebaum, Samuel, B., *Demineralization by Ion Exchange in Water Treatment and Chemical Processing of Other Liquids*, Academic Press, New York, 1968, p. 165. With permission.

Flow Scheme: Two-Bed, Strong Acid, Weak Base, Degassifier

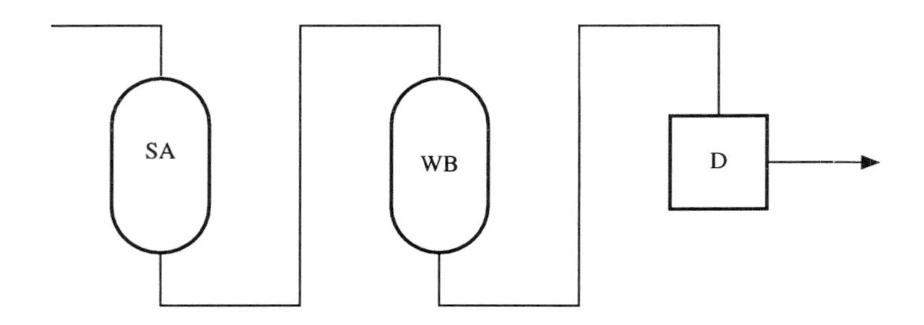

Silica unchanged; carbon dioxide removed; 10–20 µmho/cm³.

Flow Scheme: Two-Bed, Silica Removal

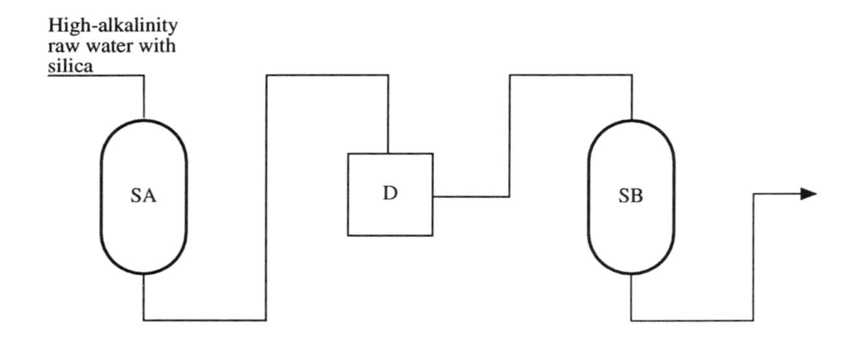

This scheme will reduce silica to 0.02–0.10 ppm; water quality 5–15 µmho/cm³.

Flow Scheme: Three-Bed (Optional Decarbonator/Degassifier)

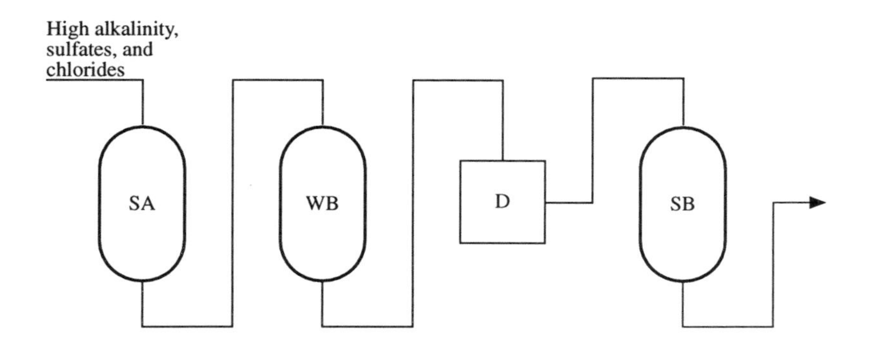

This system saves sodium hydroxide (caustic soda) regenerant with waters high in total mineral acidity (sulfates and chlorides). It also allows omission of the carbonator. The caustic soda is passed counterflow through the strong-base and weak-base beds, when the effluent conductivity of the weak-base resin rises. The strong-base bed will be only partially exhausted at that time; therefore, it has sufficient excess capacity to remove the carbonic acid economically, even if the decarbonator is omitted. This scheme also avoids early silica breakthrough.

Flow Scheme: Four-Bed with Primary Weak-Base and Secondary Strong-Base Resins

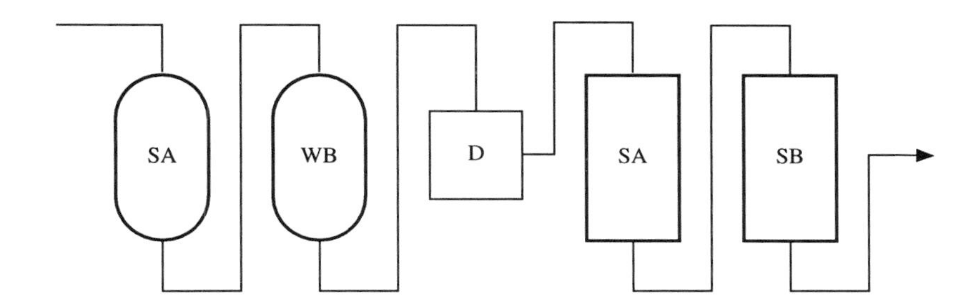

This system reduces the electrolyte concentration to 0.2–1.0 mg/L and the silica to 0.02–0.1 mg/L. The use of primary and secondary cation resins saves acid regenerant. The system is suited to waters with high percentages of sodium and low percentages of alkalinity, which cause greater sodium leakage. Weak-base resin also saves caustic soda with waters high in percentage of total mineral acidity. The secondary units, used only for polishing, can be smaller than the primary units.

Flow Scheme: Four-Bed with Primary and Secondary Strong-Base Resins

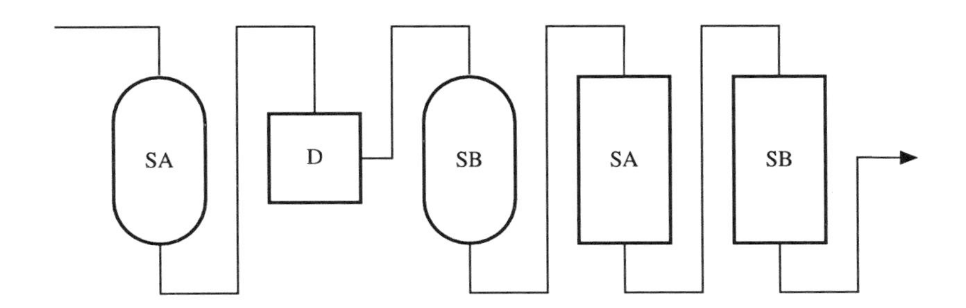

This system saves as much acid as does the previous system, but the strong-base resins in both the primary and secondary stages reduce the silica to 0.01–0.05 mg/L. The system also avoids organic fouling of the secondary resins.

Flow Scheme: Two-Bed/Four-Bed

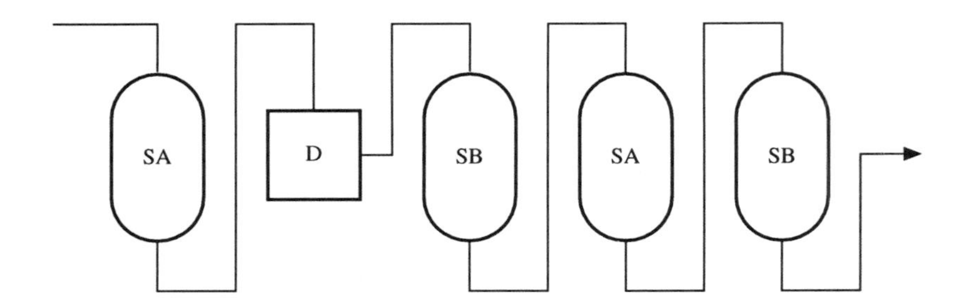

This system design is used when the output of demineralized make-up water must be increased for a short period. Then the primary and secondary stages of the four-bed are run in parallel as a large two-bed system. The secondary units are made the same size as the primary, because during regeneration of the primary pair the secondary pair handles the full load. Some two-bed/four-bed plants have been so designed that the freshly regenerated primary pair is always switched to the secondary position by opening certain valves in the piping interconnections. This is intended to improve the effluent quality, but because organic fouling of the secondary units has interfered with obtaining this objective, it is best to keep the primary units always in a primary position, so as to protect the secondary anion resin from such fouling. This scheme treats high alkalinity, high sodium, raw water to high purity (1–5 µmhos, 0.01–0.05 mg/L silica).

Flow Scheme: Mixed-Bed

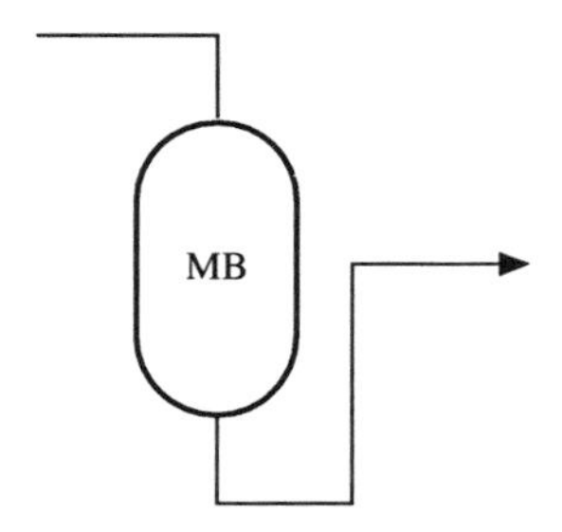

This system is used in smaller plants to treat low-TDS waters requiring high purity (1–5 µmhos, 0.01–0.05 mg/L silica) and to save investment cost, but the practice entails higher operating costs. The resin capacity in mixed beds is usually assumed to be 80–85% of the same resins in two-bed systems, so more acid and caustic soda are required for regeneration.

Flow Scheme: Cation Bed, Decarbonator, Mixed Bed

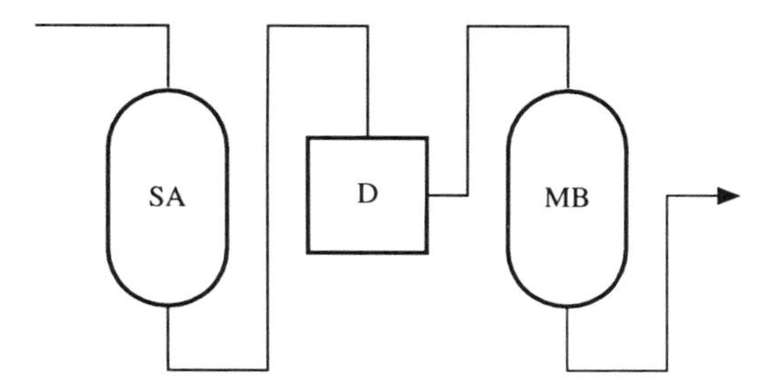

This system reduces the amount of acid and caustic soda needed compared with mixed-bed alone for waters high in alkalinity.

Flow Scheme: Two-Bed with Weak-Base and Mixed-Bed Exchangers

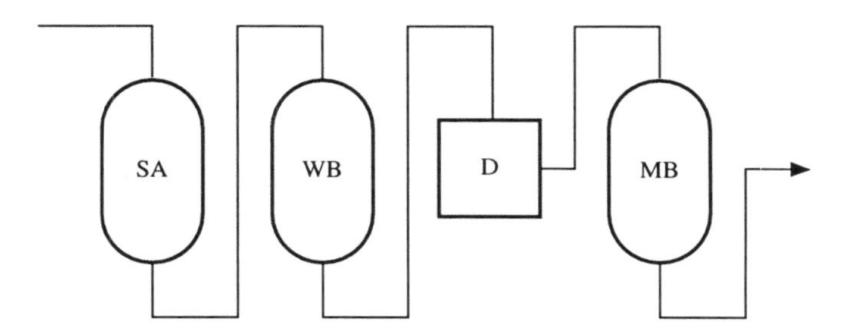

This scheme produces an effluent quality of 0.04–0.10 mg/L. It saves as much acid as the SA-D-MB system and is applicable to waters high in total mineral acidity.

Flow Scheme: Two-Bed with Strong-Base Exchanger and Mixed Bed

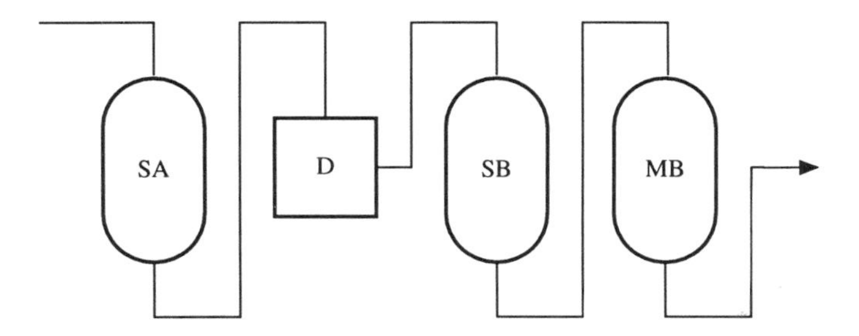

This scheme treats high-alkalinity, high-TDS water to less than 1 μmho/cm^3, silica to 0.01–0.05 ppm. The strong base exchanger protects the mixed bed from organic fouling. The mixed bed is regenerated infrequently.

Flow Scheme: Two-Bed with Weak-Acid Polishing

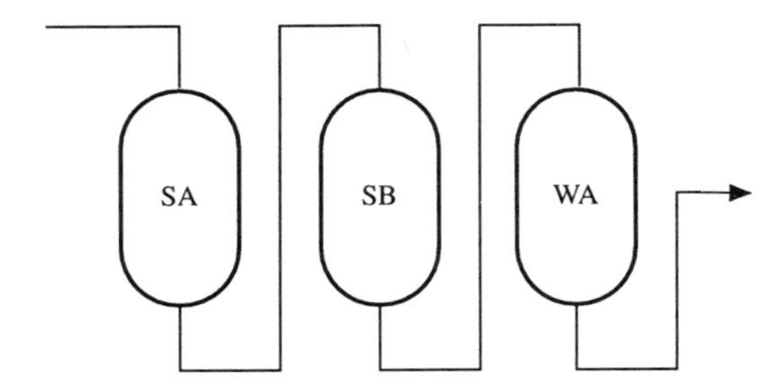

This system is used for high sodium waters. The weak-acid exchanger has a high affinity for NaOH and thus any sodium that leaks from the strong-base exchanger. The spent acid from regenerating the strong-acid exchanger is used to regenerate the weak-acid exchanger.

Alkalinity Reduction Flow Schemes

Bicarbonate alkalinity decomposes under heat, releasing carbon dioxide, which causes corrosion problems. The following flow schemes show how ion exchange can be used to dealkalize waters. The cation exchangers in the schemes are operated in the hydrogen form to convert the bicarbonate salts to carbonic acid.

The second flow scheme uses a split-stream approach in which one exchanger is operated in the hydrogen cycle only and the other in the hydrogen cycle and then blended. Either hydrochloric acid or sulfuric acid is used to regenerate the hydrogen cycle. Sodium chloride is used to regenerate the sodium cycle exchanger. In most cases a deaerator is used to remove carbon dioxide.

In the fourth flow scheme all sulfates and bicarbonate alkalinity are replaced by chloride, providing an expensive way to alkalize water when high chlorides are not a problem.

Flow Scheme: Weak-Acid Exchanger and Decarbonator

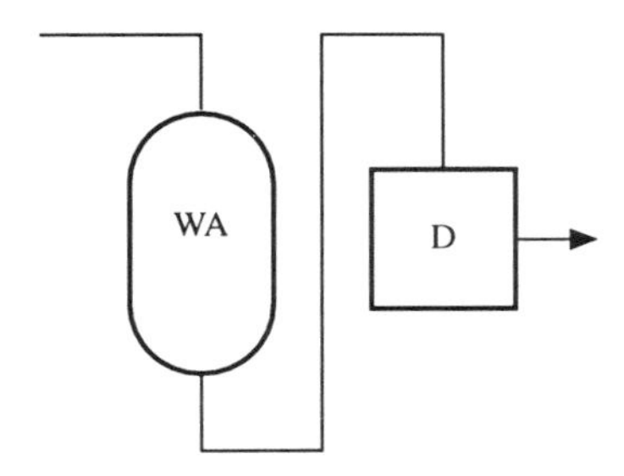

Flow Scheme: Two Strong-Acid Exchangers and Decarbonator

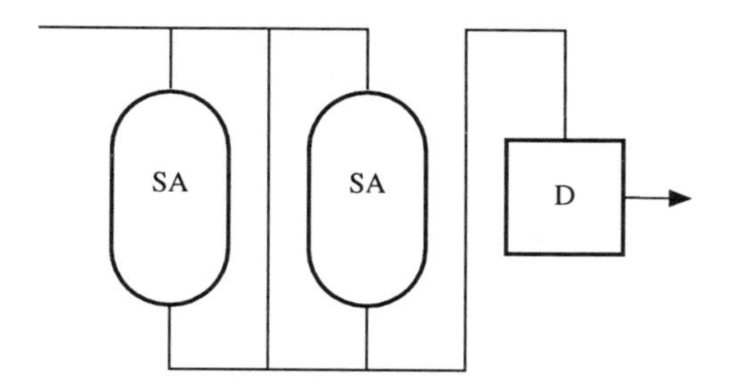

Flow Scheme: Strong-Acid Exchanger and Decarbonator with Sodium Hydroxide Addition

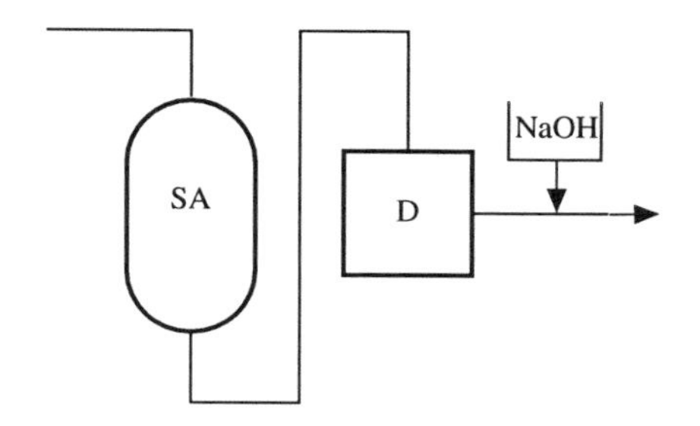

High-alkalinity, low-FMA (free mineral acidity) waters.

Flow Scheme: Dealkalizing

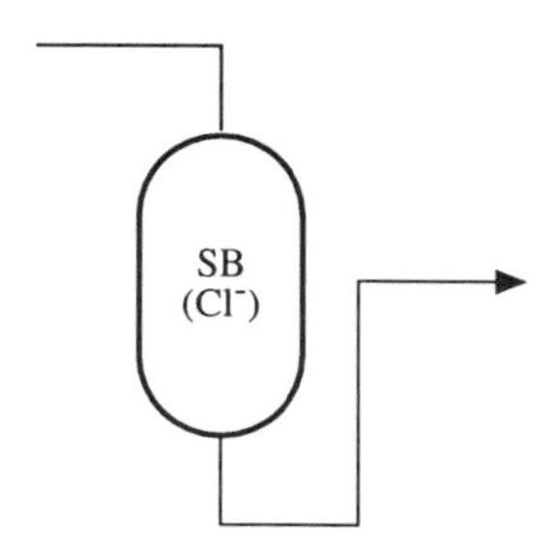

DESIGN

Formulate what you want the ion exchange design to do. To do this, the designer must determine the impurities that must be removed, which depends on the water source — well, river, process, or potable — and the use for which the treated water is intended.

Compile manufacturer's data. Analyze the water source. At a nominal cost, the resin manufacturer will provide you with all of the engineering data you need to select a resin and design the equipment. Reviewing the resin spec sheets will allow you to order the analyses necessary to complete your design. A number of the data sheets are provided for your review with permission from Rohm and Haas. Knowing the chemical quality of the water that you are to produce, the next step is to have pH, temperature, TDS, and complete cation-anion analysis of the influent or raw water. If you are producing demineralized water, you will need to know the carbon dioxide and silica concentrations. For industrial wastewaters and recovery operations, suspended solids, oil and grease (solvent extractables), organics, and a total cation-anion balance should be performed.

A good rule of thumb is that when the TDS levels of water to be deionized by exchange exceed 1000 mg/L the economics of regeneration may be questionable. Use the 1000 mg/L level as a flag. Remember, the entire picture must be considered. High TDS solutions greater than 3000 mg/L can cause self-regeneration problems.

Gather operational data. Pressure drop across the bed is a function of flow rate and water temperature. Graphs are provided by the resin manufacturer that allow computation of pressure drop as a function of flow rate and water temperature (i.e., psi/ft of bed). Design of the system must also include the pressure drop from friction losses in the piping to size pumps. A least-cost design incorporates the lowest level of regeneration and the highest service flow rate (see Figure 3-6).

Bed expansion is a function of flow rate, temperature, viscosity, and density of the resin. Typical bed expansion values for a particular resin at various water temperatures are provided by the resin manufacturer. To ensure proper cleaning of the bed, allow volume in the ion exchange vessel to accommodate 50–75% bed expansion. The effect of temperature on bed expansion is important. Design for the lowest water temperature expected (see Figure 3-13).

The effect of flow rate versus exchange capacity is also provided by the resin manufacturer. Expect breakthrough capacity to drop as service flow rates increase. Operational exchange capacity and leakage are a function of flow rate and temperature. The hardness-to-alkalinity ratio, total cations, and percent sodium must be considered.

Select the resin(s) to be used and the form in which you are to operate (a review of Chapter 3 and the flow sheets provided here will prove helpful). In selecting the resin, other considerations are breakthrough capacity for a particular operation, resistance to fracture, resistance to oxidation, and temperature operating range (see Chapter 4).

Determine pretreatment requirements. Ion exchange resins have a high affinity for solids and, therefore, act as an excellent filtration medium. But, using an exchange resin as a suspended solid filter is *not* recommended (resin prices vary between $60.00 and $400.00 per cubic foot). The resin manufacturing data sheets address a resin's resistance to a particular insult, e.g., oxidants, iron, temperature, and organics. Ferric iron forms insoluble hydroxides that can coat the resin. Ferrous iron will foul the resin as it oxidizes to the ferric state. However, if the water contains no dissolved oxygen and if iron or manganese is in the divalent state, then ion exchange is an excellent way to remove iron and manganese effectively. If the iron and manganese are in the trivalent state, or if the water contains dissolved oxygen or oxidants, or if oxygen cannot be completely excluded from the system, both iron and manganese must be removed prior to softening.

Anion exchangers are often used to remove organics. If organic removal is not in your design, pretreat for organics with activated carbon — it is more efficient and economical.

Oil and grease are often encountered when treating rinse waters from plating operations. Oil will coat the resin, thereby reducing its exchange capacity. Look for oil in industrial applications and pretreat accordingly.

Select the regenerant. Use the resin manufacturer's literature to balance regeneration level (lb of regenerant/ft^3) against capacity (Kgr/ft^3). Efficiency of regeneration and leakage are important considerations (Table 3-6).

Select regeneration contact time. Consult the manufacturer's literature. For most resins, a maximum flow rate of 1 gpm/ft^3 is recommended for most industrial softening applications. Lower rates are typically used in domestic softening. Manufacturers provide curves to select optimum regenerant levels.

Select rinse requirements. Following regeneration, the excess salt or acid must be removed (rinsed) before the service cycle can begin. In softening applications, rinsing is usually accomplished with raw water. In deionization applications, rinsing is completed with decationized water. Rinsing is usually performed in two steps; the first step is at the same rate as regeneration to displace the regerant from the resin's void spaces. After two bed volumes of rinse water have passed through the bed, the flow rate is increased to 1.5 gpm/ft^3. Rinsing is terminated at some set point — in softening, usually when the chloride content of the regenerant is 25 ppm greater than the chloride level of the raw water.

Designing an ion exchange unit.

- Compute grains per gallon as $CaCO_3$
- Compute kilograins of ions as $CaCO_3$ to be removed each day
- Select resin and calculate capacity, or determine capacity in the lab
- Calculate resin volume
- Compute regeneration requirements; select regenerant
- Calculate regeneration time
- Calculate flow rates — backwash, regeneration, rinse, service
- Determine unit dimensions
- Recheck flow rates
- Select number of units

EXAMPLE PROBLEM: DESIGN OF DEIONIZATION UNIT

Produce 100,000 gal/day of deionized water with a concentration of 10 μmhos. The 10 μmhos is equal to the leakage through the deionization units. No blending is possible. Experience has shown that for this particular manufacturer's resins, the breakthrough capacity of the cation resin is 18 Kgr/ft^3 and that for the anion resin the breakthrough capacity is 16 Kgr/ft^3. The manufacturer's literature specifies 70 gal of rinse water per cubic foot of resin. Analysis of the feed water has shown no suspended solids or oil and grease. A bar graph of the feed water has shown the following species present in the water:

$$CO_2 = 8 \text{ mg/L as } CO_2$$

$$Ca(HCO_3)_2 = 240 \text{ mg/L as } CaCO_3$$

$$Mg(HCO_3)_2 = 60 \text{ mg/L as } CaCO_3$$

$$MgSO_4 = 75 \text{ mg/L as } CaCO_3$$

$$Na_2SO_4 = 15 \text{ mg/L as } CaCO_3$$

$$NaCl = 110 \text{ mg/L as } CaCO_3$$

$$\text{Total} = 500 \text{ mg/L as } CaCO_3$$

- **Compute gr/gal as $CaCO_3$:**

$$500 \text{ mg/L total solids} \times \frac{1 \text{ gr/gal}}{17.12 \text{ mg/L}} = 29.21 \text{ gr/gal}$$

- **Compute Kgr/gal as $CaCO_3$ to be removed each day:**

$$29.21 \text{ gr/gal} \times 100,000 \text{ gal} \times \frac{1 \text{ Kgr}}{1000 \text{ gr}} = 2921 \text{ Kgr}$$

$$2921 \text{ Kgr} \div 18 \text{ Kgr/ft}^3 = 162.3 \text{ ft}^3 \text{ of cation resin}$$

$$2921 \text{ Kgr} \div 16 \text{ Kgr/ft}^3 = 182.6 \text{ ft}^3 \text{ of anion resin}$$

- **Compute regeneration needs:**

$$182.6 \text{ ft}^3 \times 70 \text{ gal for rinsing} = 12,780 \text{ gal for rinsing of anion resin}$$

$$12,780 \text{ gal} \times 29.21 \text{ gr/gal} \times \text{Kgr/1000 gr} = 373.3 \text{ Kgr of additional cations as } CaCO_3 \text{ to be removed}$$

$$373.3 \text{ Kgr} \div 18 \text{ Kgr} = 20.7 \text{ ft}^3 \text{ of cation resin}$$

Total cation resin required = 162.3 + 20.7 = 183 ft^3
With three units:

$$\text{Cation units: } 183 \text{ ft}^3 \div 3 \text{ units} = 61 \text{ ft}^3$$

$$\text{Anion units: } 182.6 \text{ ft}^3 \div 3 \text{ units} = 60.9 \text{ ft}^3$$

- **Compute service flow rates:**

$$100,000 \text{ gal} + 12,780 \text{ gal} = 112,780 \text{ gal}$$

With two units on line, one unit being regenerated:

$$112,780 \text{ gal}/2 = \frac{56,390 \text{ gal}}{\text{day} \cdot \text{unit}}$$

$$\frac{56,390 \text{ gal}}{\text{day} \cdot \text{unit}} \times \frac{1 \text{ day}}{1440 \text{ min}} = \frac{39.16 \text{ gal}}{\text{min} \cdot \text{unit}}$$

$$\frac{39.16 \text{ gal}}{\text{min} \cdot \text{unit}} \times \frac{1 \text{ ft}^2 \cdot \text{min}}{5 \text{ gal}} = 7.83 \text{ ft}^2/\text{unit}$$

A 42"-diameter tank with a 9.62 ft^2 is okay.

Resin Depth

$$61.0 \text{ ft}^3 \quad 9.62 \text{ ft}^2 = 6.34 \text{ ft deep}$$

$$60.9 \text{ ft}^2 \quad 9.62 \text{ ft}^2 = 6.33 \text{ ft deep}$$

Plating Wastewater Example Problem

Resin Selection

Strong-Acid Cation Exchanger (sodium form)

Density g/cc, 1.32
Shipping wt., 54 lb/ft^3
Effective size, 0.54
Moisture, 42%
pH range, 0–14
Swelling Na^+ - H^+, 3–7%
High capacity
High physical stability
Excellent bead integrity

Cations to be removed as $CaCO_3$

Ca^{++} = 1.0 mg/L as $CaCO_3$
Mg^{++} = 1.25

Ni^{++} = 11.0
K^{+} = 25.0
Sn^{++} = 1.0
Pb^{++} = 6.0
Cu^{++} = 6.0

Solids
Total suspended solids 200 mg/L
Total cations, 50 mg/L as $CaCO_3$

$$50\frac{\text{mg}}{\text{L}}\times\frac{1\text{ gr/gal}}{17.12\text{ mg/L}}=3\text{ gr/gal}$$

$$3\text{ gr/gal}\times 50{,}400\text{ gal/day}=151{,}200\text{ gr/day}=151.2\text{ Kgr/day}$$

$$\text{Max capacity}=2.1\text{ eq/L}=45.8\text{ Kgr/ft}^3$$

Operating capacity at 15 lb NaCl/ft^3 = 32 Kgr/ft^3
Resin required for 1-day regeneration = 5 ft^3, 151.2 Kgr/day ÷ 32 Kgr/ft^3

Check hydraulics, use 30" bed depth (2.5 ft)

$$\frac{5\text{ ft}^3}{2.5\text{ ft}}=2.0\text{ ft}^2\text{ of resin bed area}$$

$$\frac{35\text{ gpm}}{2\text{ ft}^2}=17.5\text{ gpm/ft}^2\text{ — high }\left(4-10\text{ gpm/ft}^2\right)$$

Use 6 gpm/ft^2

$$35\text{ gpm}/6\text{ gpm/ft}^2=6\text{ ft}^2$$

$$\frac{\pi D^2}{4}=6$$

$$D^2=\frac{4\times 6}{\pi}$$

$$D=3\text{ ft}^2$$

Area = 7.07 ft^2
Resin volume = (2.5) (7.07) = 17.67 → 18 ft^3

Check space velocity 2 gpm/ft^3

$$\frac{35\text{ gpm}}{18\text{ ft}^3}=2\text{ gpm/ft}^3\text{ OK}$$

$$\frac{35\text{ gpm}}{7.07\text{ ft}^2}=5\text{ gpm/ft}^2\text{ OK}$$

Allow for 75% bed expansion during backwash.
2.5 × 1.75 = 4.4 ft → 5 ft, to include 6 in. of underbedding
Use a 5-ft-high, 3-ft-diameter straight-wall vessel.

Regeneration frequency: 32 $Kgr/ft^3 \times 18\ ft^3$ resin = 576,000 gr

$$\frac{576{,}000\ \text{gr}}{151{,}200\ \text{gr/day}} = 3.8\ \text{days}$$

Increase resin volume to allow 4 days = 19 ft^3 of resin
Use same tank — 19 ft^3 resin
Use the following scheme:
Monitor breakthrough in Vessel #1, breakthrough — off-line.
Send flow to 2 during regeneration, after regeneration — in series.
Use #2 as polish unit — have to regenerate very infrequently.
Regenerate with NaCl (10%) at 15 lb/ft^3.
Regeneration flow rate = 1 gpm/ft^3 = 19–20 gpm
Regeneration level — 15 $lb/ft^3 \times 19\ ft^3$ = 285 lb NaCl

$$\text{at 10\% NaCl}\quad \frac{285\ \text{lb}}{0.897\ \text{lb/gal}} = 320\ \text{gal}$$

$$\frac{320\ \text{gal}}{19\ \text{gpm}} = 17\ \text{min}$$

Need contact time of 30 min.

$$Q = \frac{320\ \text{gal}}{30\ \text{min}} = 10\ \text{gpm}$$

Check

$$10\ \text{gpm} \times 30\ \text{min} \times 0.897\ \text{lb/gal} = 269\ \text{lb} \quad \text{— short}$$

Use 11 gpm = 296 OK

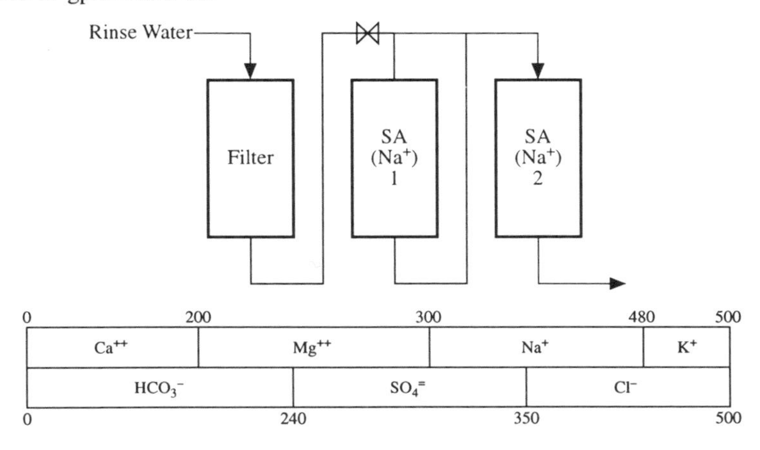

All above are as $CaCO_3$.

1. Draw a diagram of the ion exchange system.
2. Show flows through unit.
3. Show cycle in which each unit operates.

EXAMPLE PROBLEM

Starting with the following water you are to produce a finished water that has no hardness, only chlorides as anions, and an acidity as $CaCO_3$ that is 10% of the TDS. You may use two strong-acid cation units and one weak-base anion unit to produce 10,000 gal/day of water.

SOLUTION

$$Ca(HCO_3)_2 = 200 \text{ mg/L as } CaCO_3$$

$$Mg(HCO_3)_2 = 40 \text{ mg/L as } CaCO_3$$

$$MgSO_4 = 60 \text{ mg/L as } CaCO_3$$

$$Na_2SO_4 = 50 \text{ mg/L as } CaCO_3$$

$$NaCl = 130 \text{ mg/L as } CaCO_3$$

$$KCl = 20 \text{ mg/L as } CaCO_3$$

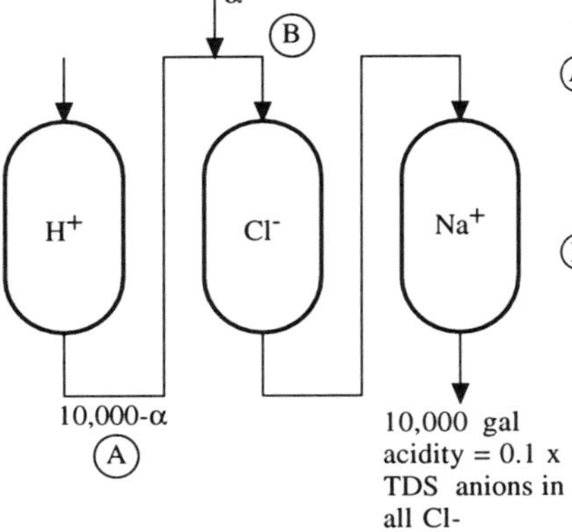

Ⓐ $H_2CO_3 = 240$ mg/L
Acidity = 110 mg/L H_2SO_4 + 150 mg/L HCl
= 260 mg/L total
Let x = TDS

Ⓑ Alkalinity-acidity
$-0.1 \times x \times 10{,}000 \text{ gal} = (10{,}000 - \alpha) \times -260 \text{ mg/L} + 240 \text{ mg/L} \times \alpha$
$-1000\ x = -2{,}600{,}000 + 500\ \alpha$
TDS
$10{,}000 \times x = 260 \text{ mg/L}\ (10{,}000 - \alpha) + 500\ \alpha$

Must use this scheme because weak-base unit will only work when pH is below 7.

$$10{,}000\ x = 2{,}600{,}000 + 240\ \alpha$$

$$10 \times (-1000\ x = -2{,}600{,}000 + 500\ \alpha)$$

$$23{,}400{,}000 = 4760\ \alpha$$

$$\alpha = 4916 \text{ gal}$$

$$10{,}000 - \alpha = 5084 \text{ gal}$$

TDS

$$10{,}000\ x = 2{,}600{,}000 - 240 \times 4916$$
$$x = 142\ \text{mg/L}$$

Alkalinity-acidity

$$4916\ \text{gal} \times 240\ \text{mg/L} - 5084\ \text{gal} \times 260\ \text{mg/L} = 10{,}000\ \text{gal} \times \beta$$
$$\beta = 14.2\ \text{mg/L acidity}$$

Example Problem: Industrial Water Treatment

Starting with the following water, produce a water that has 40 mg/L of acidity as $CaCO_3$, no hardness, and only chlorides as the anions. Bicarbonate cannot replace chloride on the resin. The TDS in the finished water may not be less than 260 mg/L as $CaCO_3$. Water flow, if needed for the problem, can be 10,000 gal/day.

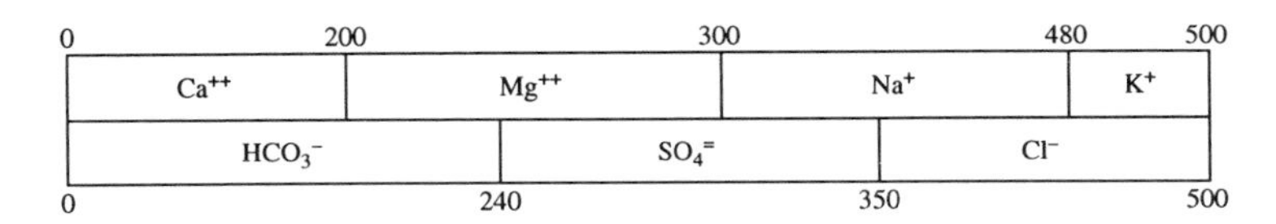

$Ca(HCO_3)_2$ = 200 mg/L as $CaCO_3$

$Mg(HCO_3)_2$ = 40 mg/L as $CaCO_3$

$MgSO_4$ = 60 mg/L as $CaCO_3$

Na_2SO_4 = 50 mg/L as $CaCO_3$

$NaCl$ = 130 mg/L as $CaCO_3$

KCl = 20 mg/L as $CaCO_3$

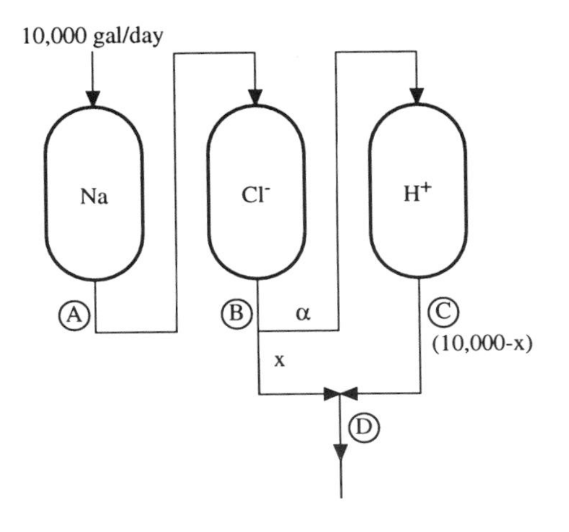

Ⓐ 240 mg/L $NaHCO_3$ as $CaCO_3$
110 mg/L Na_2SO_4 as $CaCO_3$
150 mg/L NaCl as $CaCO_3$

Ⓑ 240 mg/L $NaHCO_3$ as $CaCO_3$
260 mg/L NaCl as $CaCO_3$

Ⓒ 260 mg/L HCl as $CaCO_3$
240 mg/L H_2CO_3 as $CaCO_3$

Ⓓ 0 hardness; only Cl^-
40 mg/L acidity as $CaCO_3$
260 mg/L TDS as $CaCO_3$ min

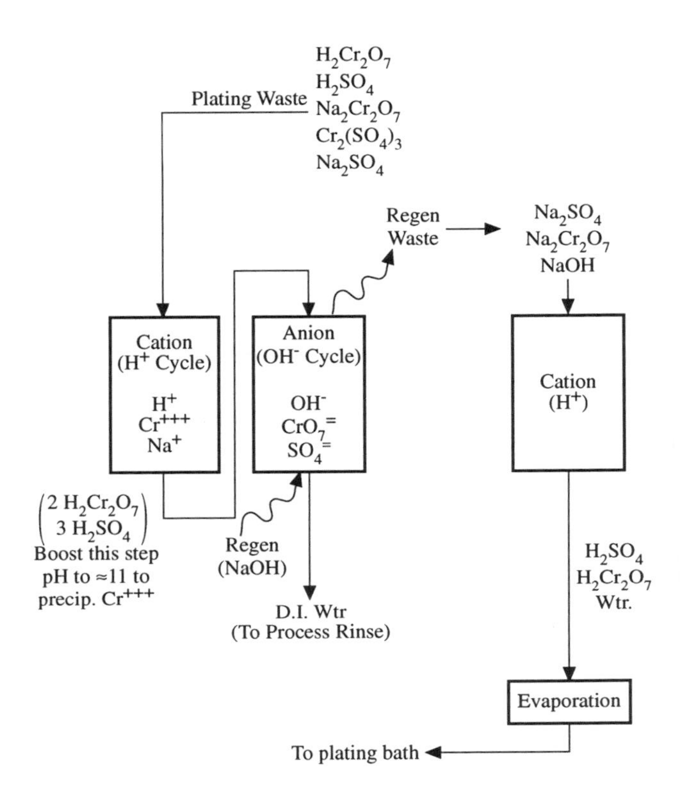

Figure 5-3 Chromium recovery scheme.

Alkalinity

$$240 \text{ mg/L} \times x + (-260 \text{ mg/L}) \times (10{,}000 - x) = 10{,}000 \times -40 \text{ mg/L}$$

$$x = 4400 \text{ gal/day}$$

NaCl

$$500 \text{ mg/L} \times 4400 \text{ gal/day} = \alpha \times 10{,}000 \text{ gal/day}$$

$$\alpha = 220 \text{ mg/L NaCl as } CaCO_3$$

TDS

$$40 \text{ mg/L HCl as } CaCO_3 + 220 \text{ mg/L NaCl as } CaCO_3$$

$$= 260 \text{ mg/L TDS as } CaCO_3$$

INDUSTRIAL FLOW SCHEMES

Nickel Removal

A simple sodium cycle cation unit will easily remove the nickel from metal plating rinse waters. However, the process itself consists of a bath at pH 4.5–5.5 containing nickel chloride and nickel sulfate salts, plus some organic brighteners and boric acid. If we were to use a hydrogen cycle system rather than a sodium

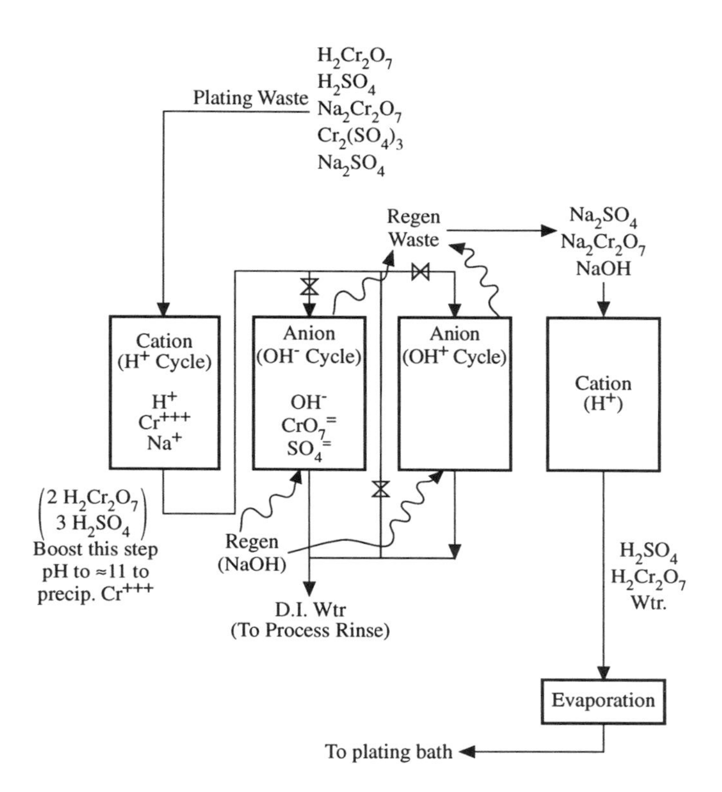

Figure 5-4 Dr. Etzel's Flip-Flop. Chromium recovery scheme.

cycle we could use a mixture of H_2SO_4 and HCl such that the regenerant wastewater had the same percentages of nickel chloride and nickel sulfate. Nickel can be added back to the bath as nickel carbonate, which will also neutralize the excess acids in the regenerant waste. The nickel carbonate-enriched regenerant wastewater thus becomes the feed to the plating bath, thus recycling the expensive and toxic nickel. This mixture is passed through an evaporator to concentrate it before returning to the bath, at which point the bicarbonate formed from the carbonate addition goes off as CO_2 into the air. Follow the hydrogen cycle unit with a hydroxyl unit and reuse the rinse waters.

Chromium Removal

Chromium plating is inefficient; that is, 15% is plated onto the parts and the remaining 85% is discharged into the rinse waters. A convention flow scheme to recover chromium is shown in Figure 5-3. Schemes similar to this are used extensively to recover chromium. One problem is that chlorides are concentrated and put back into the bath. Dr. Etzel has developed a novel way to recover chromium and remove chlorides. See Figure 5-4. Dr. Etzel uses the law that divalent ions displace monovalent ions. Any chlorides in the plating waste will

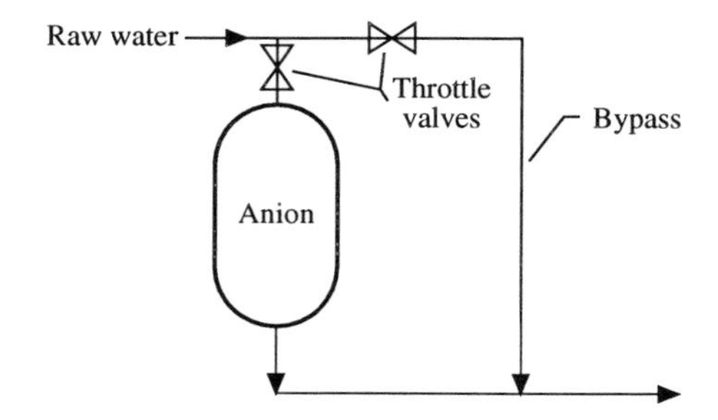

Figure 5-5 Ion exchange to feed a disinfectant.

be pushed to the second anion unit, allowing the first unit to be regenerated without chlorides. During the next cycle the second anion unit follows the cation unit and all chlorides are pushed onto the first anion unit. Flip-flop five times. On the fifth regenerate to waste. This is an inexpensive way of removing chlorides.

Note: Never load a resin higher than 0.10 lb chromium per cubic foot of resin. Any higher and you may dissolve the resin. At 0.1 lb/ft^3 expect a resin life of 1 year; at 0.5 lb/ft^3 expect 3 months; at 1.0 lb/ft^3 expect a liquid.

ION EXCHANGE AS A CHEMICAL FEEDER

The principle of ion exchange can be used to feed chemicals to a water rather than remove them. An ion exchanger operated in a given ionic form and operated in a column mode will exchange ions until the resin is exhausted and breakthrough is reached. This principle allows us to feed ions in direct proportion to the ion content of the water fed into the unit.

One example would be to feed a small amount of nitrogen and phosphorus to a wastewater stream entering biological treatment. How? Operate an ammonium cycle cation exchanger and a phosphate cycle anion exchanger set to feed, based on the ion content of the water.

Another use? To fine-tune pH using a hydrogen cycle exchanger.

If a smaller amount of material is to be fed in relation to the total ions present, bypass part of the flow (see Figure 5-5). Dr. Etzel developed a process to feed potassium dichloroisocyanurate using ion exchange. This compound breaks down into carbon dioxide, nitrogen, and chlorine, thus acting as free chlorine. This reaction takes place only in the presence of organics and is otherwise stable, even in the presence of ultraviolet light. Figure 3-4 shows the process. The potassium dichloroisocyanurate is exchanged onto the anion resin, where it is exchanged for the ions in a water, thus feeding chlorine for disinfection. The system has application in underdeveloped areas where chemical handling is not possible.

REFERENCES

Applebaum, Samuel, B., *Demineralization by Ion Exchange in Water Treatment and Chemical Processing of Other Liquids*, Academic Press, New York, 1968, p. 165.

Appendix: Tables and Conversion Factors

The tables and figures on the following pages were taken from *Engineering Manual for the Amberlite® Ion Exchange Resins*. They are reproduced by permission from Rohm and Haas Co., Philadelphia, PA.

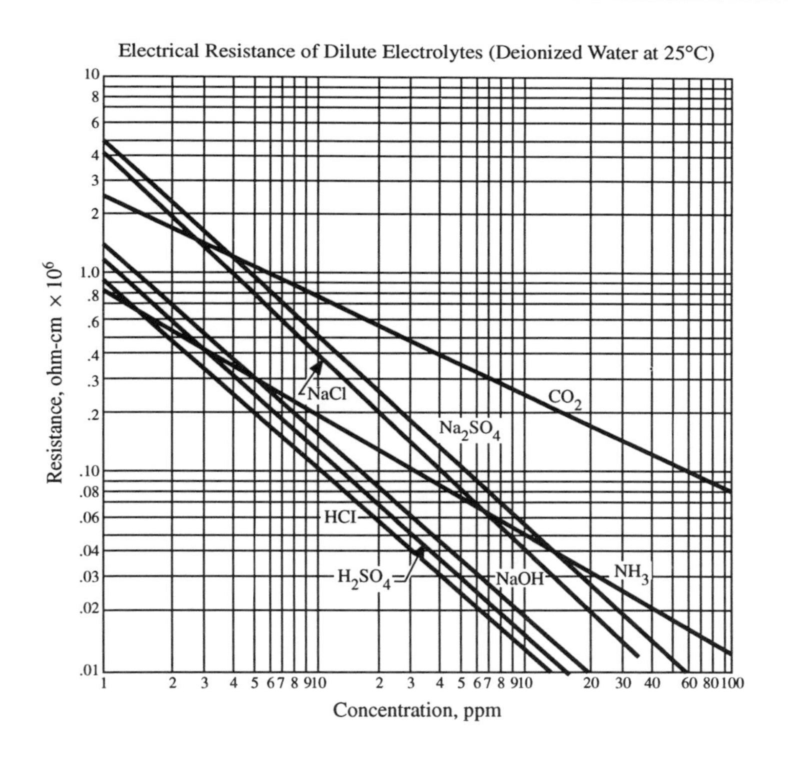

Conductivity (µmhos-cm) at 25°C	Resistivity (ohms-cm) at 25°C	Dissolved solids (ppm)	Approximate (gr/gal) as $CaCO_3$
0.056	18,000,000	0.0277	0.00164
0.059	17,000,000	0.0294	0.00170
0.063	16,000,000	0.0313	0.00181
0.067	15,000,000	0.0333	0.00193
0.072	14,000,000	0.0357	0.00211
0.077	13,000,000	0.0384	0.00222
0.084	12,000,000	0.0417	0.00240
0.091	11,000,000	0.0455	0.00263
0.100	10,000,000	0.0500	0.00292
0.111	9,000,000	0.0556	0.00322
0.125	8,000,000	0.0625	0.00368
0.143	7,000,000	0.0714	0.00415
0.167	6,000,000	0.0833	0.00485
0.200	5,000,000	0.100	0.00585
0.250	4,000,000	0.125	0.00731
0.333	3,000,000	0.167	0.00971
0.500	2,000,000	0.250	0.0146
1.00	1,000,000	0.500	0.0292
1.11	900,000	0.556	0.0322
1.25	800,000	0.625	0.0368
1.43	700,000	0.714	0.0415
1.67	600,000	0.833	0.0485
2.00	500,000	1.00	0.0585
2.50	400,000	1.25	0.0731
3.33	300,000	1.67	0.0971
5.00	200,000	2.50	0.146
10.0	100,000	5.00	0.292
11.1	90,000	5.56	0.322
12.5	80,000	6.25	0.368
14.3	70,000	7.14	0.415
16.7	60,000	8.33	0.485
20.00	50,000	10.0	0.585
25.0	40,000	12.5	0.731
33.3	30,000	16.7	0.971
50.0	20,000	25.0	1.46
100.0	10,000	50.0	2.92
111	9,000	55.6	3.22
125	8,000	62.5	3.68
143	7,000	71.4	4.15
167	6,000	83.3	4.85
200	5,000	100	5.85
250	4,000	125	7.31
333	3,000	167	9.71
500	2,000	250	14.6
1,000	1,000	500	29.2

Conductivity (µmhos-cm) at 25°C	Resistivity (ohms-cm) at 25°C	Dissolved solids (ppm)	Approximate (gr/gal) as $CaCO_3$
1,110	900	556	32.2
1,250	800	625	36.8
1,430	700	714	41.5
1,670	600	833	48.5
2,000	500	1,000	58.5
2,500	400	1,250	73.1
3,330	300	1,670	97.1
5,000	200	2,500	146
10,000	100	5,000	292

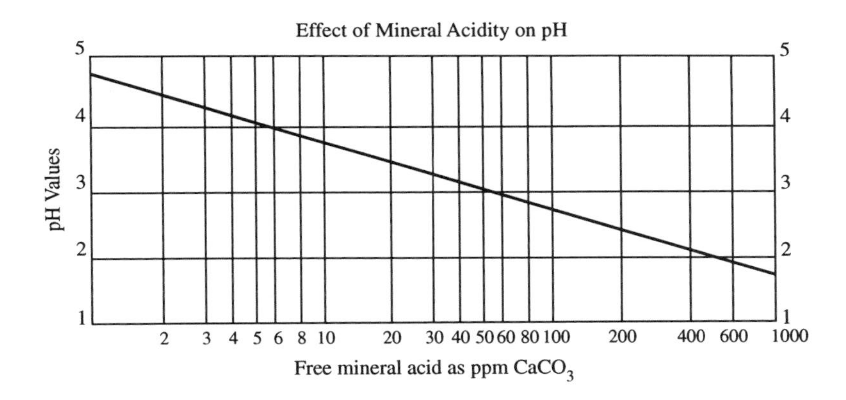

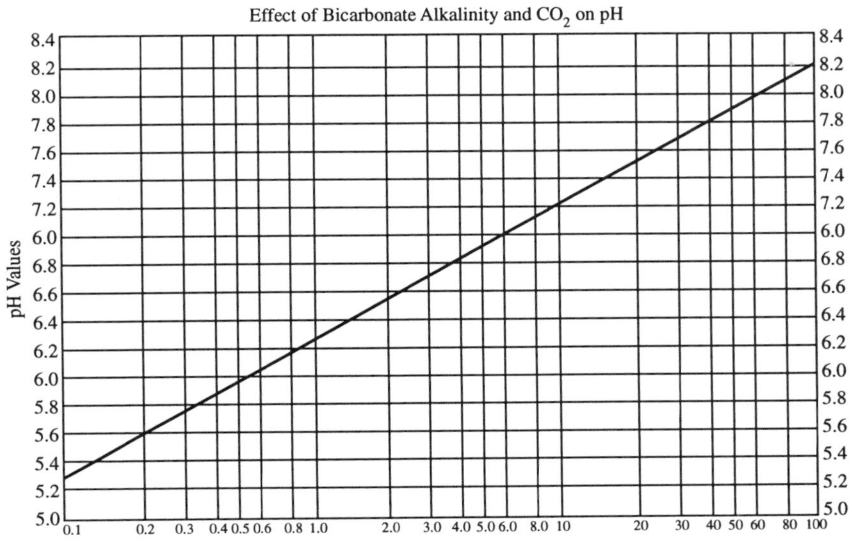

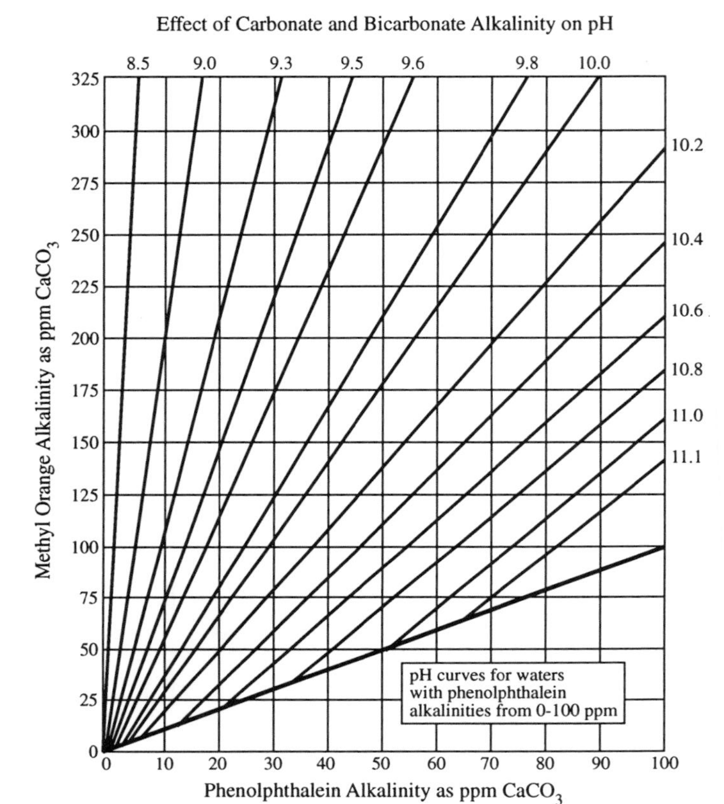

Note: pH value will also depend on the temperature of the water. The chart above is based on temperature of 20 to 25°C. As water temperature decreases, the pH value for any given combination of alkalinity forms will increase slightly above the value indicated on the chart. For example, at 5°C, actual pH will be about 0.2 units higher in the 8.5 to 9.0 pH range, about 0.3 units higher in the 9.0 to 10.0 pH range, and above pH 10 actual pH will be 0.4 to 0.6 pH units higher than indicated by the chart.

Sulfuric Acid (Aqueous Sulfuric Acid Solutions)

Specific gravity	Baumé	% H_2SO_4	Normality	g/L	lb/ft^3	lb/gal
1.0051	0.7	1	0.2051	10.05	0.6275	0.0839
1.0118	1.7	2	0.4127	20.24	1.263	0.1689
1.0184	2.6	3	.6234	30.55	1.907	0.2550
1.0250	3.5	4	.8360	41.00	2.560	0.3422
1.0317	4.5	5	1.053	51.59	3.220	0.4305
1.0385	5.4	6	1.271	62.31	3.890	0.5200
1.0453	6.3	7	1.493	73.17	4.568	0.6106
1.0522	7.2	8	1.717	84.18	5.255	0.7025
1.0591	8.1	9	1.945	95.32	5.950	0.7955
1.0661	9.0	10	2.174	106.6	6.655	0.8897
1.0731	9.9	11	2.408	118.0	7.369	0.9851
1.0802	10.8	12	2.643	129.6	8.092	1.082
1.0874	11.7	13	2.885	141.4	8.825	1.180
1.0947	12.5	14	3.126	153.3	9.567	1.279
1.1020	13.4	15	3.373	165.3	10.32	1.379
1.1094	14.3	16	3.619	177.5	11.08	1.481
1.1168	15.2	17	3.884	189.9	11.85	1.584
1.1243	16.0	18	4.127	202.4	12.63	1.689
1.1318	16.9	19	4.387	215.0	13.42	1.795
1.1394	17.7	20	4.647	227.9	14.23	1.902
1.1471	18.6	21	4.916	240.9	15.04	2.010
1.1548	19.4	22	5.181	254.1	15.86	2.120
1.1626	20.3	23	5.457	267.4	16.69	2.231
1.1704	21.1	24	5.728	280.9	17.54	2.344
1.1783	21.9	25	6.012	294.6	18.39	2.458
1.1862	22.8	26	6.289	308.4	19.25	2.574
1.1942	23.6	27	6.579	322.4	20.13	2.691
1.2023	24.4	28	6.864	336.6	21.02	2.809
1.2104	25.2	29	7.163	351.0	21.91	2.929
1.2185	26.0	30	7.455	365.6	22.82	3.051
1.2267	26.8	31	7.761	380.3	23.74	3.173
1.2349	27.6	32	8.059	395.2	24.67	3.298
1.2432	28.4	33	8.313	410.3	25.61	3.424
1.2515	29.1	34	3.676	425.5	26.56	3.551
1.2599	29.9	35	9.00	441.0	27.53	3.680
1.2684	30.7	36	9.311	456.6	28.51	3.811
1.2769	31.4	37	9.643	472.5	29.49	3.943
1.2855	32.2	38	9.961	488.5	30.49	4.077
1.2941	33.0	39	10.30	504.7	31.51	4.212
1.3028	33.7	40	10.63	521.1	32.53	4.349
1.3116	34.5	41	10.98	537.8	33.57	4.488
1.3205	35.2	42	11.31	554.6	34.62	4.628
1.3294	35.9	43	11.66	571.6	35.69	4.770
1.3384	36.7	44	12.01	588.9	36.76	4.914
1.3476	37.4	45	12.38	606.4	37.86	5.061
1.3569	38.1	46	12.73	624.2	38.97	5.209
1.3663	38.9	47	13.11	642.2	40.09	5.359
1.3758	39.6	48	13.47	660.4	41.23	5.511
1.3854	40.3	49	13.85	678.8	42.38	5.665
1.3951	41.1	50	14.22	697.6	43.55	5.821

Sulfuric Acid (Aqueous Sulfuric Acid Solutions) (continued)

Specific gravity	Baumé	% H_2SO_4	Normality	g/L	lb/ft³	lb/gal
1.4049	41.8	51	14.62	716.5	44.73	5.979
1.4148	42.5	52	15.00	735.7	45.93	6.140
1.4248	43.2	53	15.41	755.1	47.14	6.302
1.4350	44.0	54	15.80	774.9	48.37	6.467
1.4453	44.7	55	16.22	794.9	49.62	6.634
1.4557	45.4	56	16.62	815.2	50.89	6.803
1.4662	46.1	57	17.05	835.7	52.17	6.974
1.4768	46.8	58	17.46	856.5	53.47	7.148
1.4875	47.5	59	17.91	877.6	54.79	7.324
1.4983	48.2	60	18.33	899.0	56.12	7.502
1.5091	48.9	61	18.79	920.6	57.47	7.682
1.5200	49.6	62	19.22	942.4	58.83	7.865
1.5310	50.3	63	19.68	964.5	60.21	8.049
1.5421	51.0	64	20.12	986.9	61.61	8.236
1.5533	51.7	65	20.61	1010	63.03	8.426
1.5646	52.3	66	21.06	1033	64.46	8.618
1.5760	53.0	67	21.55	1056	65.92	8.812
1.5874	53.7	68	22.00	1079	67.39	9.008
1.5989	54.3	69	22.51	1103	68.87	9.207
1.6105	55.0	70	22.98	1127	70.38	9.408
1.6221	55.6	71	23.51	1152	71.90	9.611
1.6338	56.3	72	23.98	1176	73.44	9.817
1.6456	56.9	73	24.51	1201	74.99	10.02
1.6574	57.5	74	25.00	1226	76.57	10.24
1.6692	58.1	75	25.55	1252	78.15	10.45
1.6810	58.7	76	26.06	1278	79.75	10.66
1.6927	59.3	77	26.59	1303	81.37	10.88
1.7043	59.9	78	27.10	1329	82.99	11.09
1.7158	60.5	79	27.65	1355	84.62	11.31
1.7272	61.1	80	28.18	1382	86.26	11.53
1.7383	61.6	81	28.73	1408	87.90	11.75
1.7491	62.1	82	29.24	1434	89.54	11.97
1.7594	62.6	83	29.79	1460	91.16	12.19
1.7693	63.0	84	30.30	1486	92.78	12.40
1.7786	63.5	85	30.85	1512	94.38	12.62
1.7872	63.9	86	31.34	1537	95.95	12.83
1.7951	64.2	87	31.87	1562	97.49	13.03
1.8022	64.5	88	32.34	1586	99.01	13.23
1.8087	64.8	89	32.85	1610	100.5	13.43
1.8144	65.1	90	33.30	1633	101.9	13.63
1.8195	65.3	91	33.79	1656	103.4	13.82
1.8240	65.5	92	34.22	1678	104.8	14.00
1.8279	65.7	93	34.64	1700	106.1	14.19
1.8312	65.8	94	35.09	1721	107.5	14.36
1.8337	65.9	95	35.55	1742	108.7	14.54
1.8355	66.0	96	35.93	1762	110.0	14.70
1.8364	66.0	97	36.34	1781	111.2	14.87
1.8361	66.0	98	36.68	1799	112.3	15.02
1.8342	65.9	99	37.36	1816	113.4	15.15
1.8305	65.8	100	37.34	1831	114.3	15.28

Sodium Hydroxide

Specific gravity	Baumé	% NaOH	Normality	g/L	lb/ft³	lb/gal
1.0095	1.4	1	0.2524	10.10	0.6302	0.0842
1.0207	2.9	2	0.5101	20.41	1.274	0.1704
1.0318	4.5	3	0.7814	30.95	1.932	0.2583
1.0428	6.0	4	1.042	41.71	2.604	0.3481
1.0538	7.4	5	1.317	52.69	3.289	0.4397
1.0648	8.8	6	1.597	63.89	3.988	0.5332
1.0758	10.2	7	1.902	75.31	4.701	0.6284
1.0869	11.6	8	2.175	86.95	5.428	0.7256
1.0979	12.9	9	2.470	98.81	6.168	0.8246
1.1089	14.2	10	3.772	110.9	6.923	0.9254
1.1309	16.8	12	3.392	135.7	8.472	1.133
1.1530	19.2	14	4.034	161.4	10.08	1.347
1.1751	21.6	16	4.699	188.0	11.74	1.569
1.1972	23.9	18	5.387	215.5	13.45	1.798
1.2191	26.1	20	6.094	243.8	15.22	2.035
1.2411	28.2	22	6.824	273.0	17.05	2.279
1.2629	30.2	24	7.577	303.1	18.92	2.529
1.2848	32.1	26	8.349	334.0	20.85	2.788
1.3064	34.0	28	9.145	365.8	22.84	3.053
1.3279	35.8	30	9.96	398.4	24.87	3.324
1.3490	37.5	32	10.79	431.7	26.95	3.602
1.3696	39.1	34	11.64	465.7	29.07	3.886
1.3900	40.7	36	12.51	500.4	31.24	4.176
1.4101	42.2	38	13.39	535.8	33.45	4.472
1.4300	43.6	40	14.30	572.0	35.71	4.773
1.4494	45.0	42	15.22	608.7	38.00	5.080
1.4685	46.3	44	16.15	646.1	40.34	5.392
1.4873	47.5	46	17.10	684.2	42.71	5.709
1.5065	48.8	48	18.08	723.1	45.14	6.035
1.5253	49.9	50	19.07	762.7	47.61	6.364

Sodium Carbonate (Aqueous Solutions)

Specific gravity	Baumé	% Na_2CO_3	Normality	g/L	lb/ft^3	lb/gal
1.0086	1.2	1	0.1904	10.09	0.6296	0.0842
1.0190	2.7	2	0.3845	20.38	1.272	0.1701
1.0398	5.6	4	0.8979	47.59	2.596	0.3471
1.0606	8.3	6	1.201	63.64	3.973	0.5311
1.0816	10.9	8	1.633	86.53	5.402	0.7221
1.1029	13.5	10	2.081	110.3	6.885	0.9204
1.1244	16.0	12	2.545	134.9	8.423	1.126
1.1463	18.5	14	3.028	160.5	10.02	1.339

Specific gravity	Baumé	% Na_2CO_3 +10 H_2O	Normality	g/L	lb/ft^3	lb/gal
1.0086	1.2	2.70	0.1904	27.23	1.700	0.2272
1.0190	2.7	5.40	0.3847	55.02	3.435	0.4592
1.0398	5.6	10.80	0.7853	112.3	7.010	0.9370
1.0606	8.3	16.20	1.2013	171.8	10.72	1.434
1.0816	10.9	21.60	1.6335	233.6	14.58	1.949
1.1029	13.5	27.00	2.0818	297.7	18.59	2.485
1.1244	16.0	32.40	2.5475	364.3	22.74	3.040
1.1463	18.5	37.80	3.0300	433.3	27.05	3.616

Ammonium Hydroxide Solution

Specific gravity	Baumé	% NH_3	Normality	g/L	lb/ft³	lb/gal
0.9939	10.9	1	0.5836	9.939	0.6205	0.0829
0.9895	11.5	2	1.162	19.79	1.235	0.1652
0.9811	11.7	4	2.304	39.24	2.450	0.3275
0.9730	13.9	6	3.428	58.38	3.644	0.4872
0.9651	15.1	8	4.536	77.21	4.820	0.6443
0.9575	16.2	10	5.622	95.75	5.977	0.7991
0.9501	17.3	12	6.694	114.0	7.117	0.9515
0.9430	18.5	14	7.751	132.0	8.242	1.102
0.9362	19.5	16	8.796	149.8	9.351	1.250
0.9295	20.6	18	9.824	167.3	10.44	1.396
0.9229	21.7	20	10.84	184.6	11.52	1.540
0.9164	22.8	22	11.84	201.6	12.59	1.682
0.9101	23.8	24	12.82	218.4	13.64	1.823
0.9040	24.9	26	13.80	235.0	14.67	1.961
0.8980	25.9	28	14.76	251.4	15.70	2.098
0.8920	27.0	30	15.71	267.6	16.71	2.233

Hydrochloric Acid (Aqueous Hydrochloric Acid Solutions)

Specific gravity	Baumé	% HCl	Normality	g/L	lb/ft³	lb/gal
1.0032	0.5	1	.2750	10.03	0.6263	0.0837
1.0082	1.2	2	.5528	20.16	1.259	0.1683
1.0181	2.6	4	1.117	40.72	2.542	0.3399
1.0279	3.9	6	1.691	61.67	3.850	0.5147
1.0376	5.3	8	2.276	83.01	5.182	0.6927
1.0474	6.6	10	2.871	104.7	6.539	0.8741
1.0574	7.9	12	3.480	126.9	7.921	1.059
1.0675	9.2	14	4.100	149.5	9.330	1.247
1.0776	10.4	16	4.728	172.4	10.76	1.439
1.0878	11.7	18	5.370	195.8	12.22	1.634
1.0980	12.9	20	6.022	219.6	13.71	1.833
1.1083	14.2	22	6.686	243.8	15.22	2.035
1.1187	15.4	24	7.363	268.5	16.76	2.241
1.1290	16.6	26	8.049	293.5	18.32	2.450
1.1392	17.7	28	8.748	319.0	19.91	2.662
1.1493	18.8	30	9.456	344.8	21.52	2.877
1.1593	19.9	32	10.17	371.0	23.16	3.096
1.1691	21.0	34	10.90	397.5	24.81	3.317
1.1789	22.0	36	11.64	424.4	26.49	3.542
1.1885	23.0	38	12.38	451.6	28.19	3.769
1.1980	24.0	40	13.14	479.2	29.92	3.999

Sodium Chloride Solutions (15°C or 60°F)

Baumé	Specific gravity	% NaCl	lb/gal of brine		Gallons-water/ Gallons-brine	Pounds-NaCl/ Gallon-water
			NaCl	Water		
0.6	1.004	0.528	0.044	8.318	0.999	0.044
1.1	1.007	1.056	0.089	8.297	0.996	0.089
1.6	1.011	1.584	0.133	8.287	0.995	0.134
2.1	1.015	2.112	0.178	8.275	0.993	0.179
2.7	1.019	2.640	1.224	8.262	0.992	0.226
3.3	1.023	3.167	1.270	8.250	0.990	0.273
3.7	1.026	3.695	1.316	8.229	0.988	0.320
4.2	1.030	4.223	1.362	8.216	0.987	0.367
4.8	1.034	4.751	1.409	8.202	0.985	0.415
5.3	1.038	5.279	1.456	8.188	0.983	0.464
5.8	1.042	5.807	1.503	8.175	0.982	0.512
6.4	1.046	6.335	0.552	8.159	0.980	0.563
6.9	1.050	6.863	0.600	8.144	0.978	0.614
7.4	1.054	7.391	0.649	8.129	0.976	0.665
7.9	1.058	7.919	0.698	8.113	0.974	0.716
8.5	1.062	8.446	0.747	8.097	0.972	0.768
9.0	1.066	8.974	0.797	8.081	0.970	0.821
9.5	1.070	9.502	0.847	8.064	0.968	0.875
10.0	1.074	10.030	0.897	8.047	0.966	0.928
10.5	1.078	10.558	0.948	8.030	0.964	0.983
11.0	1.082	11.086	0.999	8.012	0.962	1.039
11.5	1.086	11.614	1.050	7.994	0.960	1.094
12.0	1.090	12.142	1.102	7.976	0.958	1.151
12.5	1.094	12.670	1.154	7.957	0.955	1.208
12.9	1.098	13.198	1.207	7.937	0.953	1.266
13.4	1.102	13.725	1.260	7.918	0.951	1.325
13.9	1.106	14.253	1.313	7.898	0.948	1.385
14.4	1.110	14.781	1.366	7.878	0.946	1.444
14.8	1.114	15.309	1.420	7.858	0.943	1.505
15.3	1.118	15.837	1.475	7.836	0.941	1.568
15.8	1.122	16.365	1.529	7.815	0.938	1.629
16.2	1.126	16.893	1.584	7.794	0.936	1.692
16.7	1.130	17.421	1.639	7.772	0.933	1.756
17.2	1.135	17.949	1.697	7.755	0.931	1.822
17.7	1.139	18.477	1.753	7.733	0.929	1.888
18.1	1.143	19.004	1.809	7.710	0.926	1.954
18.6	1.147	19.532	1.866	7.686	0.923	2.022
19.1	1.152	20.060	1.925	7.669	0.921	2.091
19.6	1.156	20.588	1.982	7.645	0.918	2.159
20.0	1.160	21.116	2.040	7.620	0.915	2.229
20.4	1.164	21.644	2.098	7.596	0.912	2.300
21.0	1.169	22.172	2.158	7.577	0.910	2.372
21.4	1.173	22.700	2.218	7.551	0.907	2.446
21.9	1.178	23.228	2.279	7.531	0.904	2.520
22.0*	1.179	23.310	2.288	7.528	0.904	2.531
22.3	1.182	23.755	2.338	7.506	0.901	2.594
22.7	1.186	24.283	2.398	7.479	0.898	2.670
23.3	1.191	24.811	2.459	7.460	0.896	2.745
23.7	1.195	25.339	2.522	7.430	0.892	2.827
24.2	1.200	25.867	2.585	7.409	0.890	2.906
24.4	1.202	26.131	2.616	7.394	0.888	2.947
24.6	1.204	26.395	2.647	7.380	0.886	2.987

* Eutectic point.

Ion Exchange Resin Conversion Tables

Volume

	Cubic feet	Gallons (U.S.)	Gallons (Imp.)	Liters	Cubic meters
1 Cubic foot	1	7.48	6.23	28.3	0.0283
1 Gallon (U.S.)	0.134	1	0.833	3.785	0.003785
1 Gallon (Imp.)	0.161	1.2	1	4.55	0.00455
1 Liter	0.0353	0.264	0.220	1	0.001
1 Cubic meter	35.3	264	220	1000	1

Mass

	Pounds	Grams	Kilograms	Grains	Kilograins
1 Pound	1	453.6	0.4536	7000	7
1 Gram	0.0022	1	0.001	15.43	0.01543
1 Kilogram	2.205	1000	1	15430	15.43
1 Grain	0.000143	0.0648	0.0000648	1	0.001
1 Kilograin	0.143	64.8	0.0648	1000	1

Density

	lb/ft^3	g/L	lb/gal (U.S.)	lb/gal (Imp.)
1 lb/ft^3	1	16	0.134	0.160
1 g/L	0.0624	1	0.00834	0.010
1 lb/gal (U.S.)	7.48	120	1	1.2
1 lb/gal (Imp.)	6.24	100	0.834	1

Concentration (Expressed as $CaCO_3$)

	ppm	meq/L	gr/gal (U.S.)	gr/gal (Imp.)	French degree of hardness	German degree of hardness
1 ppm	1	0.02	0.0585	0.0704	0.1	0.056
1 meq/L	50	1	2.9	3.48	5.0	2.8
1 gr/gal (U.S.)	17.1	0.34	1	1.20	1.71	0.96
1 gr/gal (Imp.)	14.2	0.28	0.833	1	1.42	0.80
1 French degree	10	0.20	0.585	0.704	1	0.56
1 German degree	17.9	0.36	1.05	1.26	1.78	1

Flow Rate[a]

	Bed volumes/min.	gpm (U.S.)/ft^3	gpm (Imp.)/ft^3	lb $H_2O/min/ft^3$
1 bed volume/min	1	7.48	6.24	62.4
1 gpm (U.S.)/ft^3	0.134	1	0.833	8.33
1 gpm (Imp.)/ft^3	0.161	1.20	1	10
1 lb $H_2O/min/ft^3$	0.016	0.12	0.10	1

[a] To convert flow rate per volume to flow rate per unit area, multiply flow rate per unit volume by resin volume and divide by cross-sectional area.

Capacity[a] and Regeneration Level

	meq/mL	lb-eq/ft³	Kgr (as $CaCO_3$) ft³	g CaCO/L	g $CaCO_3$/L
1 meq/mL	1	0.0624	21.8	28	50
1 lb-eq/ft³	16.0	1	349	449	801
1 Kgr (CaCO³)/ft³	0.0459	0.00286	1	1.28	2.29
1 g CaO/L	0.0357	0.00223	0.779	1	1.79
1 g $CaCO_3$/L	0.0200	0.00125	0.436	0.560	1

[a] Capacity on a dry-weight basis may be calculated as follows:

$$\text{g-meq/g of dry resin} = 6{,}240 \times \frac{\text{g-meq/mL}}{\text{Wet density in lb/ft}^3 \times \%\ \text{Solids}}$$

Screen Equivalents

U.S. standard			Tyler standard			British standard		
Sieve no.	Opening mm	Opening in.	Meshes per inch	Opening mm	Opening in.	Meshes per inch	Opening mm	Opening in.
12	1.68	0.0661	10	1.65	0.065	10	1.68	0.0660
14	1.41	0.0555	12	1.40	0.055	12	1.40	0.0553
16	1.19	0.0469	14	1.17	0.046	14	1.20	0.0474
18	1.00	0.0394	16	0.991	0.039	16	1.00	0.0395
20	0.84	0.0331	20	0.833	0.0328	18	0.853	0.0336
25	0.71	0.0280	24	0.701	0.0276	22	0.699	0.0275
30	0.59	0.0232	28	0.589	0.0232	25	0.599	0.0236
35	0.50	0.0197	32	0.495	0.0195	30	0.500	0.0197
40	0.42	0.0165	35	0.417	0.0164	36	0.422	0.0166
45	0.35	0.0138	42	0.351	0.0138	44	0.353	0.0139
50	0.297	0.0117	48	0.295	0.0116	52	0.295	0.0116
60	0.250	0.0098	60	0.246	0.0097	60	0.251	0.0099
70	0.210	0.0083	65	0.208	0.0082	72	0.211	0.0083
80	0.177	0.0070	80	0.175	0.0069	85	0.178	0.007
100	0.149	0.0059	100	0.147	0.0058	100	0.152	0.006
200	0.074	0.0029	200	0.074	0.0029	200	0.076	0.003
325	0.044	0.0017	325	0.043	0.0017	240	0.066	0.0026

Source: *Amberlite Ion Exchange Resins Laboratory Guide*, Rohm and Haas Company. With permission.

Rates of Flow (Corresponding French, British, and American Units)

Values of one unit	m^2/hr	m^3/s	L/s	1000 m^3/day	ft^3/s	ft^3/min	U.S. gal/min	U.S. MGD	U.K. gal/min	U.K. MGD
m^3/hr	1	278×10^{-6}	0.2778	0.024	9.81×10^{-3}	0.588	4.403	6.34×10^{-3}	3.667	5.28×10^{-3}
m^3/s	3600	1	1000	86.4	35.30	2118	15852	22.82	13198	19.00
L/s	3.6	0.001	1	0.0864	35.3×10^{-3}	2.118	15.85	22.8×10^{-3}	13.198	19×10^{-3}
1000 m^3/day	41.67	11.6×10^{-3}	11.575	1	0.4085	24.5	183.47	0.264	152.85	0.220
ft^3/s (= cfs = cusec)	102	28.3×10^{-3}	28.317	2.448	1	60	449	0.647	374	0.538
ft^3/min (= cfm)	1.70	472.10^{-6}	0.472	0.0408	0.0167	1	7.48	0.0108	6.235	8.98×10^{-3}
U.S. gal/min (U.S. gpm)	0.2271	6.3×10^{-5}	0.0631	5.45×10^{-3}	2.223×10^{-3}	0.1336	1	1.44×10^{-3}	0.833	1.20×10^{-3}
U.S. MGD	157.7	43.8×10^{-3}	43.80	3.785	1.546	92.80	694	1	578	0.832
U.K. gal/min (U.K. gpm)	0.2728	7.58×10^{-5}	0.0758	6.54×10^{-3}	2.674×10^{-3}	0.1605	1.201	1.73×10^{-3}	1	1.44×10^{-3}
U.K. MGD	189.42	52.6×10^{-3}	52.61	4.545	1.857	111.4	834	1.20	694	1

Souce: *Principles of Industrial Water Treatment*, Drew Chemical Corporation, App. A-16, p.275. With permission.

Atomic Numbers and Atomic Masses

Actinium	Ac	89	227.0278	Mercury	Hg	80	200.59
Aluminum	Al	13	26.98154	Molybdenum	Mo	42	95.94
Americium	Am	95	(243)	Neodymium	Nd	60	144.24
Antimony	Sb	51	121.75	Neon	Ne	10	20.179
Argon	Ar	18	39.948	Neptunium	Np	93	237.0482
Arsenic	As	33	74.9216	Nickel	Ni	28	58.70
Astatine	At	85	(210)	Niobium	Nb	41	92.9064
Barium	Ba	56	137.33	Nitrogen	N	7	14.0067
Berkelium	Bk	97	(247)	Nobelium	No	102	(259)
Beryllium	Be	4	9.01218	Osmium	Os	76	190.2
Bismuth	Bi	83	208.9804	Oxygen	O	8	15.9994
Boron	B	5	10.81	Palladium	Pd	46	106.4
Bromine	Br	35	79.904	Phosphorous	P	15	30.97376
Cadmium	Cd	48	112.41	Platinum	Pt	78	195.09
Calcium	Ca	20	40.08	Plutonium	Pu	94	(244)
Californium	Cf	98	(251)	Polonium	Po	84	(209)
Carbon	C	6	12.011	Potassium	K	19	39.0983
Cerium	Ce	58	140.12	Praseodymium	Pr	59	140.9077
Cesium	Cs	55	132.9054	Promethium	Pm	61	(145)
Chlorine	Cl	17	35.453	Protactinium	Pa	91	231.0389
Chromium	Cr	24	51.996	Radium	Ra	88	226.0254
Cobalt	Co	27	58.9332	Radon	Rn	86	(222)
Copper	Cu	29	63.546	Rhenium	Re	75	186.207
Curium	Cm	96	(247)	Rhodium	Rh	45	102.9055
Dysprosium	Dy	66	162.50	Rubidium	Rb	37	85.4678
Einsteinium	Es	99	(254)	Ruthenium	Ru	44	101.07
Erbium	Er	68	167.26	Samarium	Sm	62	150.4
Europium	Eu	63	151.96	Scandium	Sc	21	44.9559
Fermium	Fm	100	(257)	Selenium	Se	34	78.96
Fluorine	F	9	18.99840	Silicon	Si	14	28.0855
Francium	Fr	87	(223)	Silver	Ag	47	107.868
Gadolinium	Gd	64	157.25	Sodium	Na	11	22.98977
Gallium	Ga	31	69.72	Strontium	Sr	38	87.62
Germanium	Ge	32	72.59	Sulfur	S	16	32.06
Gold	Au	79	196.9665	Tantalum	Ta	73	180.9479
Hafnium	Hf	72	178.49	Technetium	Tc	43	(97)
Helium	He	2	4.00260	Tellurium	Te	52	127.60
Holmium	Ho	67	164.9304	Terbium	Tb	65	158.9254
Hydrogen	H	1	1.0079	Thallium	Tl	81	204.37
Indium	In	49	114.82	Thorium	Th	90	232.0381
Iodine	I	53	125.9045	Thulium	Tm	69	168.9342
Iridium	Ir	77	192.22	Tin	Sn	50	118.69
Iron	Fe	26	55.847	Titanium	Ti	22	47.90
Krypton	Kr	36	83.80	Tungsten	W	74	183.85
Lanthanum	La	57	138.9055	Uranium	U	92	238.029
Lawrencium	Lr	103	(260)	Vanadium	V	23	50.9414
Lead	Pb	82	207.2	Xenon	Xe	54	131.30
Lithium	Li	3	6.941	Ytterbium	Yb	70	173.04
Lutetium	Lu	71	174.97	Yttrium	Y	39	88.9059
Magnesium	Mg	12	24.305	Zinc	Zn	30	65.38
Manganese	Mn	25	54.9380	Zirconium	Zr	40	91.22
Mendelevium	Md	101	(258)				

Common Water Treatment Chemicals

Substance	Formula	Atomic or mol. wt.	Equivalent wt.	Conversion factor using mol. wt. of $CaCO_3$ as 100	
				Substance to $CaCO_3$ equivalent	$CaCO_3$ equivalent to substance
Aluminum	Al	27.0	9.0	5.56	0.18
Aluminum chloride	$AlCl_3$	133.	44.4	1.13	0.89
Aluminum chloride	$AlCl_3 \cdot 6H_2O$	241.	80.5	0.62	1.61
Aluminum hydrate	$Al(OH)_3$	78.0	26.0	1.92	0.52
Aluminum sulfate	$Al_2(SO_4)_3 \cdot 18H_2O$	666.4	111.1	0.45	2.22
Aluminum sulfate	$Al_2(SO_4)_3$(anhydrous)	342.1	57.0	0.88	1.14
Alumina	Al_2O_3	101.9	17.0	2.94	0.34
Sodium aluminate	$Na_2Al_2O_4$	163.9	27.8	1.80	0.55
Alum ammonium	$Al_2(SO_4)_3(NH_4)_2SO_4 \cdot 24H_2O$	906.6	151.1	0.33	3.02
Alum potassium	$Al_2(SO_4)_3K_2SO_4 \cdot 24H_2O$	948.8	156.1	0.32	3.12
Ammonia	NH_3	17.0	17.0	2.94	0.34
Ammonium (ion)	NH_4	18.0	18.0	2.78	0.86
Ammonium chloride	NH_4Cl	53.5	53.5	0.93	1.07
Ammonium hydroxide	NH_4OH	35.1	35.1	1.42	0.70
Ammonium sulfate	$(NH_4)_2SO_4$	132.	66.1	0.76	1.32
Barium	Ba	137.4	68.7	0.73	1.37
Barium carbonate	$BaCO_3$	197.4	98.7	0.58	1.97
Barium chloride	$BaCl_2 \cdot 2H_2O$	244.3	122.2	0.41	2.44
Barium hydroxide	$Ba(OH)_2$	171.	85.7	0.59	1.71
Barium nitrate	$Ba(NO_3)_2$	261.3	130.6	0.38	2.60
Barium oxide	BaO	153.	76.7	0.65	1.53
Barium sulfate	$BaSO_4$	233.4	116.7	0.43	2.33
Calcium	Ca	40.1	20.0	2.50	0.40

Common Water Treatment Chemicals (continued)

				Conversion factor using mol. wt. of $CaCO_3$ as 100	
Substance	Formula	Atomic or mol. wt.	Equivalent wt.	Substance to $CaCO_3$ equivalent	$CaCO_3$ equivalent to substance
Calcium bicarbonate	$Ca(HCO_3)_2$	162.1	81.1	0.62	1.62
Calcium carbonate	$CaCO_3$	100.08	50.1	1.00	1.00
Calcium chloride	$CaCl_2$	111.0	55.5	0.90	1.11
Calcium hydroxide	$Ca(OH)_2$	74.1	37.1	1.35	0.74
Calcium hypochlorite	$Ca(ClO)_2$	143.1	35.8	0.70	1.43
Calcium nitrate	$Ca(NO_3)_2$	164.1	82.1	0.61	1.64
Calcium oxide	CaO	56.1	28.0	1.79	0.56
Calcium phosphate	$Ca_3(PO_4)_2$	310.3	51.7	0.97	1.03
Calcium sulfate	$CaSO_4$ (anhydrous)	136.1	68.1	0.74	1.36
Calcium sulfate	$CaSO_4 \cdot 2H_2O$ (gypsum)	172.2	86.1	0.58	1.72
Carbon	C	12.0	3.00	16.67	0.06
Chlorine (ion)	Cl	35.5	35.5	1.41	0.71
Copper (cupric)	Cu	63.6	31.8	1.57	0.64
Copper sulfate (cupric)	$CuSO_4$	160.	80.0	0.63	1.60
Copper sulfate (cupric)	$CuSO_4 \cdot 5H_2O$	250.	125.	0.40	2.50
Iron (ferrous)	Fe''	55.8	27.9	1.79	0.56
Iron (ferric)	Fe'''	55.8	18.6	2.69	0.37
Ferrous carbonate	$FeCO_3$	116.	57.9	0.86	1.16
Ferrous hydroxide	$Fe(OH)_2$	89.9	44.9	1.11	0.90
Ferrous oxide	FeO	71.8	35.9	1.39	0.72
Ferrous sulfate	$FeSO_4$ (anhydrous)	151.9	76.0	0.66	1.52
Ferrous sulfate	$FeSO_4 \cdot 7H_2O$	278.0	139.0	0.36	2.78
Ferrous sulfate	$FeSO_4$ (anhydrous)	151.9	151.9	oxidation	

Ferric chloride	$FeCl_3$	162.	54.1	0.93	1.08
Ferric chloride	$FeCl_3 \cdot 6H_2O$	270.	90.1	0.56	1.80
Ferric hydroxide	$Fe(OH)_3$	107.	35.6	1.41	0.71
Ferric oxide	Fe_2O_3	160.	26.6	1.88	0.53
Ferric sulfate (ferrisul)	$Fe_2(SO_4)_3$	399.9	66.7	0.75	1.33
Ferrous or ferric	Fe or Fe	55.8	55.8	oxidation	
Fluorine	F	19.0	19.0	2.63	0.38
Hydrogen (ion)	H	1.01	1.01	50.0	0.02
Iodine	I	127.	127.	0.40	2.54
Lead	Pb	207.	104.	0.48	2.08
Magnesium	Mg	24.3	12.2	4.10	0.24
Magnesium bicarbonate	$Mg(HCO_3)_2$	146.3	73.2	0.68	1.46
Magnesium carbonate	$MgCO_3$	84.3	42.2	1.19	0.84
Magnesium chloride	$MgCl_2$	95.2	47.6	1.05	0.95
Magnesium hydroxide	$Mg(OH)_2$	58.3	29.2	1.71	0.58
Magnesium nitrate	$Mg(NO_3)_2$	40.3	20.2	2.48	0.40
Magnesium oxide	MgO	148.3	74.2	0.67	1.48
Magnesium phosphate	$Mg_3(PO_4)_2$	262.9	43.8	1.14	0.88
Magnesium sulfate	$MgSO_4$	120.4	60.2	0.83	1.20
Manganese (manganous)	Mn''	54.9	27.5	1.83	0.55
Manganese (manganic)	Mn'''	54.9	18.3	2.73	0.37
Manganese chloride	$MnCl_2$	125.8	62.9	0.80	1.26
Manganese dioxide	MnO_2	86.9	21.7	2.30	0.43
Manganese hydroxide	$Mn(OH)_2$	89.0	44.4	1.13	0.89
Manganic oxide	Mn_2O_3	158.	26.3	1.90	0.53
Manganous oxide	MnO	70.9	35.5	1.41	0.71
Nitrogen (valence 3)	N'''	14.0	4.67	10.7	0.09

Common Water Treatment Chemicals (continued)

Substance	Formula	Atomic or mol. wt.	Equivalent wt.	Conversion factor using mol. wt. of $CaCO_3$ as 100	
				Substance to $CaCO_3$ equivalent	$CaCO_3$ equivalent to substance
Nitrogen (valence 5)	N′′′′′	14.0	2.80	17.9	0.06
Oxygen	O	16.0	8.00	6.25	0.16
Phosphorus (valence 3)	P′′′	31.0	10.3	4.85	0.21
Phosphorus (valence 5)	P′′′′′	31.0	6.20	8.06	0.12
Potassium	K	39.1	39.1	1.28	0.78
Potassium carbonate	K_2CO_3	138.	69.1	0.72	1.38
Potassium chloride	KCl	74.6	74.6	0.67	1.49
Potassium hydroxide	KOH	56.1	56.1	0.89	1.12
Silver chloride	AgCl	143.3	143.3	0.35	2.87
Silver Nitrate	$AgNO_3$	169.9	169.9	0.29	3.40
Silica	SiO_2	60.1	30.0	0.83	0.60
Silicon	Si	28.1	7.03	7.11	0.14
Sodium	Na	23.0	28.0	2.18	0.46
Sodium bicarbonate	$NaHCO_3$	84.0	84.0	0.60	1.68
Sodium bisulfate	$NaHSO_4$	120.0	120.0	0.42	2.40
Sodium bisulfite	$NaHSO_3$	104.0	104.0	0.48	2.08
Sodium carbonate (anhydrous)	Na_2CO_3	106.	53.0	0.94	1.06
Sodium carbonate	$Na_2CO_3 \cdot 10H_2O$	286.	143.	0.35	2.86
Sodium chloride	NaCl	58.5	58.5	0.85	1.17
Sodium hypochlorite	NaClO	74.5	37.3	0.67	1.49
Sodium hydroxide	NaOH	40.0	40.0	1.25	0.80
Sodium nitrate	$NaNO_3$	85.0	85.0	0.59	1.70

Sodium nitrite	$NaNO_2$	69.0	34.5	0.73	1.38
Sodium oxide	Na_2O	62.0	31.0	1.61	0.62
Trisodium phosphate	$Na_3PO_4 \cdot 12H_2O$ (18.7% P_2O_5)	380.2	126.7	0.40	2.53
Trisodium phosphate (anhydrous)	Na_3PO_4(43.2% P_2O_5)	164.0	54.7	0.91	1.09
Disodium phosphate	$Na_2HPO_4 \cdot 12H_2O$ (19.8% P_2O_5)	358.2	119.4	0.42	2.39
Disodium phosphate (anhydrous)	Na_2HPO_4 (50% P_2O_5)	142.0	47.3	1.06	0.95
Monosodium phosphate	$NaH_2PO_4 \cdot H_2O$ (51.4% P_2O_5)	138.1	46.0	1.09	0.92
Monosodium phosphate (anhydrous)	NaH_2PO_4 (59.1% P_2O_5)	120.0	40.0	1.25	0.80
Meta-phosphate (anhydrous)	$NaPO_3$ (69% P_2O_5)	102.0	34.0	1.47	0.68
Sodium sulfate (anhydrous)	Na_2SO_4	142.1	71.0	0.70	1.42
Sodium sulfate	$Na_2SO_4 \cdot 10H_2O$	322.1	161.1	0.31	3.22
Sodium thiosulfate	$Na_2S_2O_3$	158.1	158.1	0.63	1.59
Sodium tetrathionate	$Na_2S_4O_6$	270.2	135.1	0.37	2.71
Sodium sulfite	Na_2SO_3	126.1	63.0	0.79	1.26
Sulfur (valence 2)	S″	32.1	16.0	3.13	0.32
Sulfur (valence 4)	S′′′′	32.1	8.02	6.25	0.16
Sulfur (valence 6)	S′′′′′′	32.1	5.34	9.36	0.11
Sulfur dioxide	SO_2	64.1	32.0		
Tin	Sn	119.			
Water	H_2O	18.0	9.00	5.56	0.18
Zinc	Zn	65.4	32.7	1.53	0.65
	Acid Radicals				
Bicarbonate	HCO_3	61.0	61.0	0.82	1.22
Carbonate	CO_3	60.0	30.0	0.83[a]	.60
Carbon dioxide	CO_2	44.0	44.0	1.14	.44

Common Water Treatment Chemicals (continued)

Substance	Formula	Atomic or mol. wt.	Equivalent wt.	Conversion factor using mol. wt. of $CaCO_3$ as 100	
				Substance to $CaCO_3$ equivalent	$CaCO_3$ equivalent to substance
Chloride	Cl	35.5	35.5	1.41	.71
Iodide	I	126.9	126.9	0.39	2.54
Nitrate	NO_3	62.0	62.0	0.81	1.24
Hydroxide	OH	17.0	17.0	2.94	0.34
Phosphate	PO_4	95.0	31.7	1.58	0.63
Phosphorous oxide	P_2O_5	142.0	23.7	2.11	0.47
Sulfide	S	32.1	16.0	3.13	0.32
Sulfate	SO_4	96.1	48.0	1.04	0.96
Sulfur trioxide	SO_3	80.1	40.0	1.25	0.80
Acids					
Hydrogen	H	1.0	1.0	50.00	0.02
Acetic acid	$HC_2H_3O_2$	60.1	60.1	0.83	1.20
Carbonic acid	H_2CO_3	62.0	31.0	1.61	0.62
Hydrochloric acid	HCl	36.5	36.5	1.37	0.73
Nitric acid	HNO_3	63.0	63.0	0.79	1.26
Phosphoric acid	H_3PO_4	98.0	32.7	1.53	0.65
Sulfurous acid	H_2SO_3	82.1	41.1	1.22	0.82
Sulfuric acid	H_2SO_4	98.1	49.0	1.02	0.98
Hydrogen sulfide	H_2S	34.1	17.05	2.93	0.34
Manganous acid	H_2MnO_2	104.9	52.5	0.95	1.05

[a] In ion exchange reactions, it is assumed carbonate reacts as the monovalent ion.

Source: *Principles of Industrial Water Treatment*, Drew Chemical Corporation. Reprinted with permission of the copyright owner, Ashland, Inc.

Index

I

S

T